# BEST OF NIKOLA TESLA

My Inventions: the Autobiography Of Nikola Tesla & Other Works



SANAGE
PUBLISHING HOUSE

Paperback: 978-939492434-5



Printed by: Manipal Technologies Limited, Manipal

Sanage Publishing House LLP
Mumbai, India

sanagepublishing@gmail.com

**Nikola Tesla** (1856 – 7 January 1943) was a Serbian-American inventor, electrical engineer, mechanical engineer, and futurist best known for his contributions to the design of the modern alternating current (AC) electricity supply system.

Tesla's lifelong obsession with a wireless transmission of energy began when he showed such a transmission to the world through a teleautomatic boat guided by a remote control. He even made the discovery of terrestrial waves which proved that the earth could be used as a conductor and made to resonate at a certain frequency. He, allegedly, had also worked on a death-ray and claimed to receive signals from another planet.

Tesla obtained around 300 patents worldwide for his inventions. There are a minimum of 278 known patents issued to Tesla in 26 countries. Many inventions developed by Tesla were not put into patent protection.

# TABLE OF CONTENTS

## My Inventions Autobiography Of Nikola Tesla



## Experiments with Alternate Currents of High Potential and High Frequency



## The Problem of Increasing Human Energy

# My Inventions Autobiography Of Nikola Tesla

# I. My Early Life

The progressive development of man is vitally dependent on invention. It is the most important product of his creative brain. Its ultimate purpose is the complete mastery of mind over the material world, the harnessing of the forces of nature to human needs. This is the difficult task of the inventor who is often misunderstood and unrewarded. But he finds ample compensation in the pleasing exercises of his powers and in the knowledge of being one of that exceptionally privileged class without whom the race would have long ago perished in the bitter struggle against pitiless elements.

Speaking for myself, I have already had more than my full measure of this exquisite enjoyment, so much that for many years my life was little short of continuous rapture. I am credited with being one of the hardest workers and perhaps I am, if thought is the equivalent of labour, for I have devoted to it almost all of my waking hours. But if work is interpreted to be a definite performance in a specified time according to a rigid rule, then I may be the worst of idlers. Every effort under compulsion demands a sacrifice of life-energy. I never paid such a price. On the contrary, I have thrived on my thoughts.

In attempting to give a connected and faithful account of my activities in this series of articles which will be presented with the assistance of the Editors of the ELECTRICAL EXPERIMENTER and are chiefly address to our young men readers, I must dwell, however reluctantly, on the impressions of my youth and the circumstances and events which have been instrumental in determining my career.

Our first endeavours are purely instinctive, promptings of an imagination vivid and undisciplined. As we grow older reason asserts itself and we become more and more systematic and designing. But those early impulses, tho not immediately productive, are of the greatest moment and may shape our very destinies. Indeed, I feel now that had I understood and cultivated instead of suppressing them, I would have added substantial value to my bequest to the world. But not until I had attained manhood did, I realize that I was an inventor.

This was due to a number of causes. In the first place I had a brother who was gifted to an extraordinary degree one of those rare phenomena of mentality which biological investigation has failed to explain. His premature death left my parents disconsolate. We owned a horse which had been presented to us by a dear friend. It was a magnificent animal of Arabian breed, possessed of almost human intelligence, and was cared for and petted by the whole family, having on one occasion saved my father's life under remarkable circumstances. My father had been called one winter night to perform an urgent duty and while crossing the mountains, infested by wolves, the horse became frightened and ran away, throwing him violently to the ground. It arrived home bleeding and exhausted, but after the alarm was sounded immediately dashed off again, returning to the spot, and before the searching party were far on the way they were met by my father, who had recovered consciousness and remounted, not realizing that he had been lying in the snow for several hours. This horse was responsible for my brother's injuries from which he died. I witness the tragic scene and although fifty-six years have elapsed since, my visual impression of it has lost none of its force. The recollection of his attainments made every effort of mine seem dull in comparison.

Anything I did that was creditable merely caused my parents to feel their loss more keenly. So, I grew up with little confidence in myself. But I was far from being considered a stupid boy, if I am to judge from an incident of which I have still a strong remembrance. One day the Aldermen were passing thru a street where I was at play with other boys. The oldest of these venerable gentlemen a wealthy citizen paused to give a silver piece to each of us. Coming to me he suddenly stops and commanded, "Look in my eyes." I met his gaze, my hand outstretched to receive the much-valued coin, when, to my dismay, he said, "No, not much, you can get nothing from me, you are too smart." They used to tell a funny story about me. I had two old aunts with wrinkled faces, one of them having two teeth protruding like the tusks of an elephant which she buried in my cheek every time she kisses me. Nothing would scare me more than the prospect of being hugged by these as affectionate as unattractive relatives. It happened that while being carried in my mother's arms, they asked me who was the prettier of the two. After examining their faces intently, I answered thoughtfully, pointing to one of them, "This here is not as ugly as the other."

Then again, I was intended from my very birth for the clerical profession, and this thought constantly oppressed me. I longed to be an engineer, but my father was inflexible. He was the son of an officer who served in the army of the Great Napoleon and, in common with his brother, professor of mathematics in a prominent institution, had received a military education but, singularly enough, later embraced the clergy in which vocation he achieved eminence. He was a very erudite man, a veritable natural philosopher, poet and writer and his sermons were said to be as eloquent as those of Abraham a Sancta-Clara. He had a prodigious memory and frequently recited at length from works in several languages. He often remarked playfully that if some of the classics were lost, he could restore them. His style of writing was much admired. He penned sentences short and terse and was full of wit and satire. The humorous remarks he made were always peculiar and characteristic. Just to illustrate, I may mention one or two instances. Among the help there was a cross-eyed man called Mane, employed to do work around the farm. He was chopping wood one day. As he swung the axe my father, who stood nearby and felt very uncomfortable, cautioned him, "For God's sake, Mane, do not strike at what you are looking but at what you intend to hit." On another occasion he was taking out for a drive a friend who carelessly permitted his costly fur coat to rub on the carriage wheel. My father reminded him of it saying, "Pull in your coat, you are ruining my tire." He had the odd habit of talking to himself and would often carry on an animated conversation and indulge in heated argument, changing the tone of his voice. A casual listener might have sworn that several people were in the room.

Although I must trace to my mother's influence whatever inventiveness I possess, the training he gave me must have been helpful. It comprised all sorts of exercises as, guessing one another's thoughts, discovering the defects of some form or expression, repeating long sentences or performing mental calculations. These daily lessons were intended to strengthen memory and reason and specially to develop the critical sense and were undoubtedly very beneficial.

My mother descended from one of the oldest families in the country and a line of inventors. Both her father and grandfather originated numerous implements for household, agricultural and other uses. She was a truly great woman, of rare skill, courage and fortitude, who had braved the storms of life and passed thru many a trying experience. When she was sixteen a virulent pestilence swept the country. Her father was called away to administer the last sacraments to the dying and during his absence she went alone to the assistance of a neighbouring family who were stricken by the dread disease. All of the members, five in number, succumbed in rapid succession. She bathed, clothed and laid out the bodies, decorating them with flowers according to the custom of the country and when her father returned, he found everything ready for a Christian burial. My mother was an inventor of the first order and would, I believe, have achieved great things had she not been so remote from modern life and its multifold opportunities. She invented and constructed all kinds of tools and devices and wove the finest designs from thread which was spun by her. She even planted the seeds, raised the plants and separated the fibres herself. She worked indefatigably, from break of day till late at night, and most of the wearing apparel and furnishings of the home was the product of her hands. When she was past sixty, her fingers were still nimble enough to tie three knots in an eyelash.

There was another and still more important reason for my late awakening. In my boyhood I suffered from a peculiar affliction due to the appearance of images, often accompanied by strong flashes of light, which marred the sight of real objects and interfered with my thought and action. They were pictures of things and scenes which I had really seen, never of those I imagined. When a word was spoken to me the image of the object it designated would present itself vividly to my vision and sometimes, I was quite unable to distinguish whether what I saw was tangible or not. This caused me great discomfort and anxiety. None of the students of psychology or physiology whom I have consulted could ever explain satisfactorily these phenomena. They seem to have been unique although I was probably predisposed as I know that my brother experienced a similar trouble. The theory I have formulated is that the images were the result of a reflex action from the brain on the retina under great excitation. They certainly were not hallucinations such as are produced in diseased and anguished minds, for in other respects I was normal and composed. To give an idea of my distress, suppose that I had witnessed a funeral or some such nerve- racking spectacle. Then, inevitably, in the stillness of night, a vivid picture of the scene would thrust itself before my eyes and persist despite all my efforts to banish it. Sometimes it would even remain fixt in space tho I pushed my hand thru it. If my explanation is correct, it should be able to project on a screen the image of any object one conceives and make it visible. Such an advance would revolutionize all human relations. I am convinced that this wonder can and will be accomplished in time to come; I may add that I have devoted much thought to the solution of the problem.

To free myself of these tormenting appearances, I tried to concentrate my mind on something else I had seen, and in this way, I would often obtain temporary relief; but in order to get it I had to conjure continuously new images. It was not long before I found that I had exhausted all of those at my command; my "reel" had run out, as it were, because I had seen little of the world only objects in my home and the immediate surroundings. As I performed these mental operations for the second or third time, in order to chase the appearances from my vision, the remedy gradually lost all its force. Then I instinctively commenced to make excursions beyond the limits of the small world of which I had knowledge, and I saw new scenes. These were at first very blurred and indistinct and would flit away when I tried to concentrate my attention upon them, but by and by I succeeded in fixing them; they gained in strength and distinctness and finally assumed the concreteness of real things. I soon discovered that my best comfort was attained if I simply went on in my vision farther and farther, getting new impressions all the time, and so I began to travel of course, in my mind. Every night (and sometimes during the day), when alone, I would start on my journeys see new places, cities and countries live there, meet people and make friendships and acquaintances and, however unbelievable, it is a fact that they were just as dear to me as those in actual life and not a bit less intense in their manifestations.

This I did constantly until I was about seventeen when my thoughts turned seriously to invention. Then I observed to my delight that I could visualize with the greatest facility. I needed no models, drawings or experiments. I could picture them all as real in my mind. Thus, I have been led unconsciously to evolve what I consider a new method of materializing inventive concepts and ideas, which is radically opposite to the purely experimental and is in my opinion ever so much more expeditious and efficient. The moment one constructs a device to carry into practise a crude idea he finds himself unavoidably engross with the details and defects of the apparatus. As he goes on improving and reconstructing, his force of concentration diminishes, and he loses sight of the great underlying principle. Results may be obtained but always at the sacrifice of quality.

My method is different. I do not rush into actual work. When I get an idea, I start at once building it up in my imagination. I change the construction, make improvements and operate the device in my mind. It is absolutely immaterial to me whether I run my turbine in thought or test it in my shop. I even note if it is out of balance. There is no difference whatever, the results are the same. In this way I am able to rapidly develop and perfect a conception without touching anything. When I have gone so far as to embody in the invention every possible improvement I can think of and see no fault anywhere, I put into concrete form this final product of my brain. Invariably my device works as I conceived that it should, and the experiment comes out exactly as I planned it. In twenty years, there has not been a single exception. Why should it be otherwise? Engineering, electrical and mechanical, is positive in results. There is scarcely a subject that cannot be mathematically treated, and the effects calculated, or the results determined beforehand from the available theoretical and practical data. The carrying out into practise of a crude idea as is being generally done is, I hold, nothing but a waste of energy, money and time.

My early affliction had, however, another compensation. The incessant mental exertion developed my powers of observation and enabled me to discover a truth of great importance. I had noted that the appearance of images was always preceded by actual vision of scenes under peculiar and generally very exceptional conditions, and I was impelled on each occasion to locate the original impulse. After a while this effort grew to be almost automatic, and I gained great facility in connecting cause and effect. Soon I became aware, to my surprise, that every thought I conceived was suggested by an external impression. Not only this but all my actions were prompted in a similar way. In the course of time, it became perfectly evident to me that I was merely an automaton endowed with power of movement, responding to the stimuli of the sense organs and thinking and acting accordingly. The practical result of this was the art of Tel automatics which has been so far carried out only in an imperfect manner. Its latent possibilities will, however, be eventually shown. I have been since years planning self-controlled automata and believe that mechanisms can be produced which will act as if possessed of reason, to a limited degree, and will create a revolution in many commercial and industrial departments.

I was about twelve years old when I first succeeded in banishing an image from my vision by wilful effort, but I never had any control over the flashes of light to which I have referred. They were, perhaps, my strangest experience and inexplicable. They usually occurred when I found myself in a dangerous or distressing situation, or when I was greatly exhilarated. In some instances, I have seen all the air around me filled with tongues of living flame. Their intensity, instead of diminishing, increased with time and seemingly attained a maximum when I was about twenty-five years old. While in Paris, in 1883, a prominent French manufacturer sent me an invitation to a shooting expedition which I accepted. I had been long confined to the factory and the fresh air had a wonderfully invigorating effect on me. On my return to the city that night I felt a positive sensation that my brain had caught fire. I saw a light as tho a small sun was located in it and I past the whole night applying cold compressions to my tortured head. Finally, the flashes diminished in frequency and force, but it took more than three weeks before they wholly subsided. When a second invitation was extended to me my answer was an emphatic NO!

These luminous phenomena still manifest themselves from time to time, as when a new idea opening up possibilities strikes me, but they are no longer exciting, being of relatively small intensity. When I close my eyes, I invariably observe first, a background of very dark and uniform blue, not unlike the sky on a clear but starless night. In a few seconds this field becomes animated with innumerable scintillating flakes of green, arranged in several layers and advancing towards me. Then there appears, to the right, a beautiful pattern of two systems of parallel and closely spaced lines, at right angles to one another, in all sorts of colours with yellow-green and gold predominating. Immediately thereafter the lines grow brighter and the whole is thickly sprinkled with dots of twinkling light. This picture moves slowly across the field of vision and in about ten seconds vanishes to the left, leaving behind a ground of rather unpleasant and inert grey which quickly gives way to a billowy sea of clouds, seemingly trying to mould themselves in living shapes. It is curious that I cannot project a form into this grey until the second phase is reached. Every time, before falling asleep, images of persons or objects flit before my view. When I see them, I know that I am about to lose consciousness. If they are absent and refuse to come it means a sleepless night.

To what an extent imagination played a part in my early life I may illustrate by another odd experience. Like most children I was fond of jumping and developed an intense desire to support myself in the air. Occasionally a strong wind richly charged with oxygen blew from the mountains rendering my body as light as cork and then I would leap and float in space for a long time. It was a delightful sensation, and my disappointment was keen when later I undeceived myself.

During that period, I contracted many strange likes, dislikes and habits, some of which I can trace to external impressions while others are unaccountable. I had a violent aversion against the earrings of women but other ornaments, as bracelets, pleased me more or less according to design. The sight of a pearl would almost give me a fit, but I was fascinated with the glitter of crystals or objects with sharp edges and plane surfaces. I would not touch the hair of other people except, perhaps, at the point of a revolver. I would get a fever by looking at a peach and if a piece of camphor was anywhere in the house it caused me the keenest discomfort. Even now I am not insensible to some of these upsetting impulses. When I drop little squares of paper in a dish filled with liquid, I always sense a peculiar and awful taste in my mouth. I counted the steps in my walks and calculated the cubical contents of soup plates, coffee cups and pieces of food otherwise my meal was unenjoyable. All repeated acts or operations I performed had to be divisible by three and if I missed, I felt impelled to do it all over again, even if it took hours.

Up to the age of eight years, my character was weak and vacillating. I had neither courage nor strength to form a firm resolve. My feelings came in waves and surges and vibrated unceasingly between extremes. My wishes were of consuming force and like the heads of the hydra, they multiplied. I was oppressed by thoughts of pain in life and death and religious fear. I was swayed by superstitious belief and lived in constant dread of the spirit of evil, of ghosts and ogres and other unholy monsters of the dark. Then, all at once, there came a tremendous change which altered the course of my whole existence. Of all things I liked books the best. My father had a large library and whenever I could manage, I tried to satisfy my passion for reading. He did not permit it and would fly into a rage when he caught me in the act. He hid the candles when he found that I was reading in secret. He did not want me to spoil my eyes. But I obtained tallow, made the wicking and cast the sticks into tin forms, and every night I would bush the keyhole and the cracks and read, often till dawn, when all others slept, and my mother started on her arduous daily task. On one occasion I came across a novel entitled "Abafi" (the Son of Aba), a Serbian translation of a well-known Hungarian writer, Josika. This work somehow awakened my dormant powers of will and I began to practise self-control. At first my resolutions faded like snow in April, but in a little while I conquered my weakness and felt a pleasure I never knew before that of doing as I willed. In the course of time this vigorous mental exercise became second nature. At the outset my wishes had to be subdued but gradually desire and will grew to be identical. After years of such discipline, I gained so complete a mastery over myself that I toyed with passions which have meant destruction to some of the strongest men. At a certain age I contracted a mania for gambling which greatly worried my parents. To sit down to a game of cards was for me the quintessence of pleasure. My father led an exemplary life and could not excuse the senseless waste of time and money in which I indulged. I had a strong resolve, but my philosophy was bad. I would say to him, "I can stop whenever I please but is it worthwhile to give up that which I would purchase with the joys of Paradise?" On frequent occasions he gave vent to his anger and contempt, but my mother was different. She understood the character of men and knew that one's salvation could only be brought about thru his own efforts. One afternoon, I remember, when I had lost all my money and was craving for a game, she came to me with a roll of bills and said, "Go and enjoy yourself. The sooner you lose all we possess the better it will be. I know that you will get over it." She was right. I conquered my passion then and there and only regretted that it had not been a hundred times as strong. I not only vanquished but tore it from my heart so as not to leave even a trace of desire. Ever since that time I have been as indifferent to any form of gambling as to picking teeth.

During another period I smoked excessively, threatening to ruin my health. Then my will asserted itself and I not only stop but destroyed all inclination. Long ago I suffered from heart trouble until I discovered that it was due to the innocent cup of coffee, I consumed every morning. I discontinued at once, tho I confess it was not an easy task. In this way I checked and bridled other habits and passions and have not only preserved my life but derived an immense amount of satisfaction from what most men would consider privation and sacrifice.

After finishing the studies at the Polytechnic Institute and University I had a complete nervous breakdown and while the malady lasted, I observed many phenomena strange and unbelievable.

# II. My First Efforts At Invention

I shall dwell briefly on these extraordinary experiences, on account of their possible interest to students of psychology and physiology and also because this period of agony was of the greatest consequence on my mental development and subsequent labours. But it is indispensable to first relate the circumstances and conditions which preceded them, and in which might be found their partial explanation.

From childhood I was compelled to concentrate attention upon myself. This caused me much suffering but, to my present view, it was a blessing in disguise for it has taught me to appreciate the inestimable value of introspection in the preservation of life, as well as a means of achievement. The pressure of occupation and the incessant stream of impressions pouring into our consciousness thru all the gateways of knowledge make modern existence hazardous in many ways. Most persons are so absorbed in the contemplation of the outside world that they are wholly oblivious to what is passing on within themselves.

The premature death of millions is primarily traceable to this cause. Even among those who exercise care it is a common mistake to avoid imaginary and ignore the real dangers. And what is true of an individual also applies, more or less, to a people as a whole. Witness, in illustration, the prohibition movement. A drastic, if not unconstitutional, measure is now being put thru in this country to prevent the consumption of alcohol and yet it is a positive fact that coffee, tea, tobacco, chewing gum and other stimulants, which are freely indulged in even at the tender age, are vastly more injurious to the national body, judging from the number of those who succumb. So, for instance, during my student years I gathered from the published necrologues in Vienna, the home of coffee drinkers, that deaths from heart trouble sometimes reached sixty-seven per cent of the total. Similar observations might probably be made in cities where the consumption of tea is excessive. These delicious beverages super excite and gradually exhaust the fine fibres of the brain. They also interfere seriously with arterial circulation and should be enjoyed all the more sparingly as their deleterious effects are slow and imperceptible. Tobacco, on the other hand, is conducive to easy and pleasant thinking and detracts from the intensity and concentration necessary to all original and vigorous effort of the intellect. Chewing gum is helpful for a short while but soon drains the glandular system and inflicts irreparable damage, not to speak of the revulsion it creates. Alcohol in small quantities is an excellent tonic but is toxic in its action when absorbed in larger amounts, quite immaterial as to whether it is taken in as whiskey or produced in the stomach from sugar. But it should not be overlooked that all these are great eliminators assisting Nature, as they do, in upholding her stern but just law of the survival of the fittest. Eager reformers should also be mindful of the eternal perversity of mankind which makes the indifferent "laissez-faire" by far preferable to enforced restraint.

The truth about this is that we need stimulants to do our best work under present living conditions, and that we must exercise moderation and control our appetites and inclinations in every direction. That is what I have been doing for many years, in this way maintaining myself young in body and mind. Abstinence was not always to my liking, but I find ample reward in the agreeable experiences I am now making. Just in the hope of converting some to my precepts and convictions I will recall one or two.

A short time ago I was returning to my hotel. It was a bitter cold night, the ground slippery, and no taxi to be had. Half a block behind me followed another man, evidently as anxious as myself to get under cover. Suddenly my legs went up in the air. In the same instant there was a flash in my brain, the nerves responded, the muscles contracted, I swung thru 180 degrees and landed on my hands. I resumed my walk as tho nothing had happened when the stranger caught up with me. "How old are you?" he asked, surveying me critically. "Oh, about fifty-nine," I replied. "What of it?" "Well," said he, "I have seen a cat do this but never a man." About a month since I wanted to order new eyeglasses and went to an oculist who put me thru the usual tests. He looked at me incredulously as I read off with ease the smallest print at considerable distance. But when I told him that I was past sixty he gasped in astonishment. Friends of mine often remark that my suits fit me like gloves, but they do not know that all my clothing is made to measurements which were taken nearly 35 years ago and never changed. During this same period my weight has not varied one pound.

In this connection I may tell a funny story. One evening, in the winter of 1885, Mr. Edison, Edward H. Johnson, the President of the Edison Illuminating Company, Mr. Batchellor, Manager of the works, and myself entered a little place opposite 65 Fifth Avenue where the offices of the company were located. Someone suggested guessing weights and I was induced to step on a scale. Edison felt me all over and said: "Tesla weighs 152 lbs. to an ounce," and he guessed it exactly. Stripped I weighed 142 lbs. and that is still my weight. I whispered to Mr. Johnson: "How is it possible that Edison could guess my weight so closely?" "Well," he said, lowering his voice. "I will tell you, confidentially, but you must not say anything. He was employed for a long time in a Chicago slaughterhouse where he weighed thousands of hogs every day! That's why." My friend, the Hon. Chauncey M. Depew, tells of an Englishman on whom he sprung one of his original anecdotes and who listened with a puzzled expression but - a year later - laughed out loud. I will frankly confess it took me longer than that to appreciate Johnson's joke.

Now, my wellbeing is simply the result of a careful and measured mode of living and perhaps the most astonishing thing is that three times in my youth I was rendered by illness a hopeless physical wreck and given up by physicians. More than this, thru ignorance and light-heartedness, I got into all sorts of difficulties, dangers and scrapes from which I extricated myself as by enchantment. I was almost drowned a dozen times; was nearly boiled alive and just mist being cremated. I was entombed, lost and frozen. I had hair-breadth escapes from mad dogs, hogs, and other wild animals. I passed thru dreadful diseases and met with all kinds of odd mishaps and that I am hale, and hearty today seems like a miracle. But as I recall these incidents to my mind, I feel convinced that my preservation was not altogether accidental.

An inventor's endeavour is essentially lifesaving. Whether he harnesses forces, improves devices, or provides new comforts and conveniences, he is adding to the safety of our existence. He is also better qualified than the average individual to protect himself in peril, for he is observant and resourceful. If I had no other evidence that I was, in a measure, possessed of such qualities I would find it in these personal experiences. The reader will be able to judge for himself if I mention one or two instances. On one occasion, when about 14 years old, I wanted to scare some friends who were bathing with me. My plan was to dive under a long floating structure and slip out quietly at the other end. Swimming and diving came to me as naturally as to a duck and I was confident that I could perform the feat. Accordingly, I plunged into the water and, when out of view, turned around and proceeded rapidly towards the opposite side. Thinking that I was safely beyond the structure, I rose to the surface but to my dismay struck a beam. Of course, I quickly dived and forged ahead with rapid strokes until my breath was beginning to give out. Rising for the second time, my head came again in contact with a beam. Now I was becoming desperate. However, summoning all my energy, I made a third frantic attempt, but the result was the same. The torture of suppressed breathing was getting unendurable, my brain was reeling, and I felt myself sinking. At that moment, when my situation seemed absolutely hopeless, I experienced one of those flashes of light and the structure above me appeared before my vision. I either discerned or guest that there was a little space between the surface of the water and the boards resting on the beams and, with consciousness nearly gone, I floated up, prest my mouth close to the planks and managed to inhale a little air, unfortunately mingled with a spray of water which nearly choked me. Several times I repeated this procedure as in a dream until my heart, which was racing at a terrible rate, quieted down and I gained composure. After that I made a number of unsuccessful dives, having completely lost the sense of direction, but finally succeeded in getting out of the trap when my friends had already given me up and were fishing for my body.

That bathing season was spoiled for me thru recklessness, but I soon forgot the lesson and only two years later I fell into a worse predicament. There was a large flour mill with a dam across the river near the city where I was studying at that time. As a rule, the height of the water was only two or three inches above the dam and to swim out to it was a sport not very dangerous in which I often indulged. One day I went alone to the river to enjoy myself as usual. When I was a short distance from the masonry, however, I was horrified to observe that the water had risen and was carrying me along swiftly. I tried to get away, but it was too late. Luckily, tho, I saved myself from being swept over by taking hold of the wall with both hands. The pressure against my chest was great and I was barely able to keep my head above the surface. Not a soul was in sight and my voice was lost in the roar of the fall. Slowly and gradually I became exhausted and unable to withstand the strain longer. just as I was about to let go, to be dashed against the rocks below, I saw in a flash of light a familiar diagram illustrating the hydraulic principle that the pressure of a fluid in motion is proportionate to the area exposed, and automatically I turned on my left side. As if by magic the pressure was reduced, and I found it comparatively easy in that position to resist the force of the stream. But the danger still confronted me. I knew that sooner or later I would be carried down, as it was not possible for any help to reach me in time, even if I attracted attention. I am ambidextrous now but then I was lefthanded and had comparatively little strength in my right arm. For this reason, I did not dare to turn on the other side to rest and nothing remained but to slowly push my body along the dam. I had to get away from the mill towards which my face was turned as the current there was much swifter and deeper. It was a long and painful ordeal, and I came near to failing at its very end for I was confronted with a depression in the masonry. I managed to get over with the last ounce of my force and fell in a swoon when I reached the bank, where I was found. I had torn virtually all the skin from my left side, and it took several weeks before the fever subsided, and I was well. These are only two of many instances, but they may be sufficient to show that had it not been for the inventor's instinct I would not have lived to tell this tale.

Interested people have often asked me how and when I began to invent. This I can only answer from my present recollection in the light of which the first attempt I recall was rather ambitious for it involved the invention of an apparatus and a method. In the former I was anticipated but the latter was original. It happened in this way. One of my playmates had come into the possession of a hook and fishing-tackle which created quite an excitement in the village, and the next morning all started out to catch frogs. I was left alone and deserted owing to a quarrel with this boy. I had never seen a real hook and pictured it as something wonderful, endowed with peculiar qualities and was despairing not to be one of the parties. Urged by necessity, I somehow got hold of a piece of soft iron wire, hammered the end to a sharp point between two stones, bent it into shape, and fastened it to a strong string. I then cut a rod, gathered some bait, and went down to the brook where there were frogs in abundance. But I could not catch any and was almost discouraged when it occurred to me to dangle the empty hook in front of a frog sitting on a stump. At first, he collapsed but by and by his eyes bulged out and became bloodshot, he swelled to twice his normal size and made a vicious snap at the hook.

Immediately I pulled him up. I tried the same thing again and again and the method proved infallible. When my comrades, who in spite of their fine outfit had caught nothing, came to me they were green with envy. For a long time, I kept my secret and enjoyed the monopoly but finally yielded to the spirit of Christmas. Every boy could then do the same and the following summer brought disaster to the frogs.

In my next attempt I seem to have acted under the first instinctive impulse which later dominated me - to harness the energies of nature to the service of man. I did this thru the medium of Maybugs or June-bugs as they are called in America which were a veritable pest in that country and sometimes broke the branches of trees by the sheer weight of their bodies. The bushes were black with them. I would attach as many as four of them to a crosspiece, notably arranged on a thin spindle, and transmit the motion of the same to a large disc and so derive considerable "power." These creatures were remarkably efficient, for once they were started, they had no sense to stop and continued whirling for hours and hours and the hotter it was the harder they worked. All went well until a strange boy came to the place. He was the son of a retired officer in the Austrian Army. That urchin ate May-bugs alive and enjoyed them as tho they were the finest blue-point oysters. That disgusting sight terminated my endeavours in this promising field, and I have never since been able to touch a Maybug or any other insect for that matter.

After that, I believe, I undertook to take apart and assemble the clocks of my grandfather. In the former operation I was always successful but often failed in the latter. So, it came that he brought my work to a sudden halt in a manner not too delicate and it took thirty years before I tackled another clockwork again. Shortly thereafter I went into the manufacture of a kind of popgun which comprised a hollow tube, a piston, and two plugs of hemp. When firing the gun, the piston was prest against the stomach and the tube was pushed back quickly with both hands. The air between the plugs was compressed and raised to high temperature and one of them was expelled with a loud report. The art consisted in selecting a tube of the proper taper from the hollow stalks. I did very well with that gun, but my activities interfered with the windowpanes in our house and met with painful discouragement. If I remember rightly, I then took to carving swords from pieces of furniture which I could conveniently obtain. At that time, I was under the sway of the Serbian national poetry and full of admiration for the feats of the heroes. I used to spend hours in mowing down my enemies in the form of cornstalks which ruined the crops and netted me several spankings from my mother. Moreover, these were not of the formal kind but the genuine article.

I had all this and more behind me before I was six years old and had passed thru one year of elementary school in the village of Smiljan where I was born. At this juncture we moved to the little city of Gospic nearby. This change of residence was like a calamity to me. It almost broke my heart to part from our pigeons, chickens and sheep, and our magnificent flock of geese which used to rise to the clouds in the morning and return from the feeding grounds at sundown in battle formation, so perfect that it would have put a squadron of the best aviators of the present day to shame. In our new house I was but a prisoner, watching the strange people I saw thru the window blinds. My bashfulness was such that I would rather have faced a roaring lion than one of the city dudes who strolled about. But my hardest trial came on Sunday when I had to dress up and attend the service. There I meet with an accident, the mere thought of which made my blood curdle like sour milk for years afterwards. It was my second adventure in a church. Not long before I was entombed for a night in an old chapel on an inaccessible mountain which was visited only once a year. It was an awful experience, but this one was worse. There was a wealthy lady in town, a good but pompous woman, who used to come to the church gorgeously painted up and attired with an enormous train and attendants. One Sunday I had just finished ringing the bell in the belfry and rushed downstairs when this grand dame was sweeping out and I jumped on her train. It tore off with a ripping noise which sounded like a salvo of musketry fired by raw recruits. My father was livid with rage. He gave me a gentle slap on the cheek, the only corporal punishment he ever administered to me, but I almost feel it now. The embarrassment and confusion that followed are indescribable. I was practically ostracised until something else happened which redeemed me in the estimation of the community.

An enterprising young merchant had organized a fire department. A new fire engine was purchased, uniforms provided, and the men drilled for service and parade. The engine was, in reality, a pump to be worked by sixteen men and was beautifully painted red and black. One afternoon the official trial was prepared for, and the machine was transported to the river. The entire population turned out to witness the great spectacle. When all the speeches and ceremonies were concluded, the command was given to pump, but not a drop of water came from the nozzle. The professors and experts tried in vain to locate the trouble. The fizzle was complete when I arrived at the scene. My knowledge of the mechanism was nil, and I knew next to nothing of air pressure, but instinctively I felt for the suction hose in the water and found that it had collapsed. When I waded in the river and opened it up the water rushed forth and not a few Sunday clothes were spoiled. Archimedes running naked thru the streets of Syracuse and shouting Eureka at the top of his voice did not make a greater impression than myself. I was carried on the shoulders and was the hero of the day.

Upon settling in the city, I began a four-years' course in the so-called Normal School preparatory to my studies at the College or Real Gymnasium. During this period my boyish efforts and exploits, as well as troubles, continued. Among other things I attained the unique distinction of champion crow catcher in the country. My method of procedure was extremely simple. I would go in the forest, hide in the bushes, and imitate the call of the bird. Usually, I would get several answers and in a short while a crow would flutter down into the shrubbery near me. After that all I needed to do was to throw a piece of cardboard to distract its attention, jump up and grab it before it could extricate itself from the undergrowth. In this way I would capture as many as I desired. But on one occasion something occurred which made me respect them. I had caught a fine pair of birds and was returning home with a friend. When we left the forest, thousands of crows had gathered making a frightful racket. In a few minutes they rose in pursuit and soon enveloped us. The fun lasted until all of a sudden, I received a blow on the back of my head which knocked me down. Then they attacked me viciously. I was compelled to release the two birds and was glad to join my friend who had taken refuge in a cave.

In the schoolroom there were a few mechanical models which interested me and turned my attention to water turbines. I constructed many of these and found great pleasure in operating them. How extraordinary was my life an incident may illustrate? My uncle had no use for this kind of pastime and more than once rebuked me. I was fascinated by a description of Niagara Falls I had perused and pictured in my imagination a big wheel run by the Falls. I told my uncle that I would go to America and carry out this scheme. Thirty years later I saw my ideas carried out at Niagara and marvelled at the unfathomable mystery of the mind.

I made all kinds of other contrivances and contraptions but among these the arbalists I produced were the best. My arrows, when shot, disappeared from sight and at close range traversed a plank of pine one inch thick. Thru the continuous tightening of the bows, I developed skin on my stomach very much like that of a crocodile and I am often wondering whether it is due to this exercise that I am able even now to digest cobblestones! Nor can I pass in silence my performances with the sling which would have enabled me to give a stunning exhibit at the Hippodrome. And now I will tell of one of my feats with this antique implement of war which will strain to the utmost the credulity of the reader. I was practicing while walking with my uncle along the river. The sun was setting, the trout were playful and from time to time one would shoot up into the air, its glistening body sharply defined against a projecting rock beyond. Of course, any boy might have hit a fish under these propitious conditions, but I undertook a much more difficult task and I foretold to my uncle, to the minutest detail, what I intended doing. I was to hurl a stone to meet the fish, press its body against the rock, and cut it in two. It was no sooner said than done. My uncle looked at me almost scared out of his wits and exclaimed "Vade retro Satanas!" and it was a few days before he spoke to me again. Other records, however great, will be eclipsed but I feel that I could peacefully rest on my laurels for a thousand years.

# III. My Later Endeavours

## The Discovery of the Rotating Magnetic Field

At the age of ten I entered the Real Gymnasium which was a new and fairly well-equipped institution. In the department of physics were various models of classical scientific apparatus, electrical and mechanical. The demonstrations and experiments performed from time to time by the instructors fascinated me and were undoubtedly a powerful incentive to invention. I was also passionately fond of mathematical studies and often won the professor's praise for rapid calculation. This was due to my acquired facility of visualizing the figures and performing the operations, not in the usual intuitive manner, but as in actual life. Up to a certain degree of complexity it was absolutely the same to me whether I wrote the symbols on the board or conjured them before my mental vision. But freehand drawing, to which many hours of the course were devoted, was an annoyance I could not endure. This was rather remarkable as most of the members of the family excelled in it. Perhaps my aversion was simply due to the predilection I found in undisturbed thought. Had it not been for a few exceptionally stupid boys, who could not do anything at all, my record would have been the worst. It was a serious handicap as under the then existing educational regime, drawing being obligatory, this deficiency threatened to spoil my whole career and my father had considerable trouble in railroading me from one class to another.

In the second year at that institution, I became obsessed with the idea of producing continuous motion thru steady air pressure. The pump incident, of which I have told, had set afire my youthful imagination and impress me with the boundless abilities of a vacuum. I grew frantic in my desire to harness this inexhaustible energy but for a long time I was groping in the dark. Finally, however, my endeavours crystallized in an invention which was to enable me to achieve what no other mortal ever attempted.

Imagine a cylinder freely rotatable on two bearings and partly surrounded by a rectangular trough which fits it perfectly. The open side of the trough is closed by a partition so that the cylindrical segment within the enclosure divides the latter into two compartments entirely separated from each other by air-tight sliding joints. One of these compartments being sealed and once for all exhausted, the other remaining open, a perpetual rotation of the cylinder would result, at least, I thought so. A wooden model was constructed and fitted with infinite care and when I applied the pump on one side and actually observed that there was a tendency to turning, I was delirious with joy. Mechanical flight was the one thing I wanted to accomplish although still under the discouraging recollection of a bad fall I sustained by jumping with an umbrella from the top of a building. Every day I used to transport myself thru the air to distant regions but could not understand just how I managed to do it. Now I had something concrete, a flying machine with nothing more than a rotating shaft, flapping wings, and a vacuum of unlimited power! From that time on I made my daily aerial excursions in a vehicle of comfort and luxury as might have befitted King Solomon. It took years before I understood that the atmospheric pressure acted at right angles to the surface of the cylinder and that the slight rotary effort, I observed was due to a leak. Tho this knowledge came gradually it gave me a painful shock.

I had hardly completed my course at the Real Gymnasium when I was prostrated with a dangerous illness or rather, a score of them, and my condition became so desperate that I was given up by physicians.

During this period, I was permitted to read constantly, obtaining books from the Public Library which had been neglected and entrusted to me for classification of the works and preparation of the catalogues. One day I was handed a few volumes of new literature unlike anything I had ever read before and so captivating as to make me utterly forget my hopeless state. They were the earlier works of Mark Twain and to them might have been due the miraculous recovery which followed. Twenty-five years later, when I met Mr. Clemens and we formed a friendship between us, I told him of the experience and was amazed to see that great man of laughter burst into tears.

My studies were continued at the higher Real Gymnasium in Carlstadt, Croatia, where one of my aunts resided. She was a distinguished lady, the wife of a Colonel who was an old war-horse having participated in many battles. I never can forget the three years I passed at their home. No fortress in time of war was under a more rigid discipline. I was fed like a canary bird. All the meals were of the highest quality and deliciously prepared but short in quantity by a thousand percent. The slices of ham cut by my aunt were like tissue paper. When the Colonel would put something substantial on my plate, she would snatch it away and say excitedly to him: "Be careful, Niko is very delicate." I had a voracious appetite and suffered like Tantalus. But I lived in an atmosphere of refinement and artistic taste quite unusual for those times and conditions. The land was low and marshy and malaria fever never left me while there despite of the enormous amounts of quinin I consumed. Occasionally the river would rise and drive an army of rats into the buildings, devouring everything even to the bundles of the fierce paprika. These pests were to me a welcome diversion. I thinned their ranks by all sorts of means, which won me the unenviable distinction of ratcatcher in the community. At last, however, my course was completed, the misery ended, and I obtained the certificate of maturity which brought me to the cross-roads.

During all those years my parents never wavered in their resolve to make me embrace the clergy, the mere thought of which filled me with dread. I had become intensely interested in electricity under the stimulating influence of my Professor of Physics, who was an ingenious man and often demonstrated the principles by apparatus of his own invention. Among these I recall a device in the shape of a freely rotatable bulb, with tinfoil coatings, which was made to spin rapidly when connected to a static machine. It is impossible for me to convey an adequate idea of the intensity of feeling I experienced in witnessing his exhibitions of these mysterious phenomena. Every impression produced a thousand echoes in my mind. I wanted to know more of this wonderful force; I longed for experiment and investigation and resigned myself to the inevitable with aching heart.

Just as I was making ready for the long journey home, I received word that my father wished me to go on a shooting expedition. It was a strange request as he had been always strenuously opposed to this kind of sport. But a few days later I learned that the cholera was raging in that district and, taking advantage of an opportunity, I returned to Gospic in disregard of my parents' wishes. It is incredible how absolutely ignorant people were as to the causes of this scourge which visited the country in intervals of from fifteen to twenty years. They thought that the deadly agents were transmitted thru the air and filled it with pungent odours and smoke. In the meantime, they drank the infected water and died in heaps. I contracted the awful disease on the very day of my arrival and although surviving the crisis, I was confined to bed for nine months with scarcely any ability to move. My energy was completely exhausted and for the second time I found myself at death's door. In one of the sinking spells which was thought to be the last, my father rushed into the room. I still see his pallid face as he tried to cheer me in tones belying his assurance. "Perhaps," I said, "I may get well if you will let me study engineering." "You will go to the best technical institution in the world," he solemnly replied, and I knew that he meant it. A heavy weight was lifted from my mind, but the relief would have come too late had it not been for a marvellous cure brought about thru a bitter decoction of a peculiar bean. I came to life like another Lazarus to the utter amazement of everybody.

My father insisted that I spend a year in healthful physical outdoor exercises to which I reluctantly consented. For most of this term I roamed in the mountains, loaded with a hunter's outfit and a bundle of books, and this contact with nature made me stronger in body as well as in mind. I thought and planned and conceived many ideas almost as a rule delusive. The vision was clear enough, but the knowledge of principles was very limited. In one of my inventions, I proposed to convey letters and packages across the seas, thru a submarine tube, in spherical containers of sufficient strength to resist the hydraulic pressure. The pumping plant, intended to force the water thru the tube, was accurately figured and designed and all other particulars carefully worked out. Only one trifling detail, of no consequence, was lightly dismissed. I assumed an arbitrary velocity of the water and, what is more, took pleasure in making it high, thus arriving at a stupendous performance supported by faultless calculations. Subsequent reflections, however, on the resistance of pipes to fluid flow determined me to make this invention public property.

Another one of my projects was to construct a ring around the equator which would, of course, float freely and could be arrested in its spinning motion by reactionary forces, thus enabling travel at a rate of about one thousand miles an hour, impracticable by rail. The reader will smile. The plan was difficult of execution, I will admit, but not nearly so bad as that of a well-known New York professor, who wanted to pump the air from the torrid to the temperate zones, entirely forgetful of the fact that the Lord had provided a gigantic machine for this very purpose?

Still another scheme, far more important and attractive, was to derive power from the rotational energy of terrestrial bodies. I had discovered that objects on the earth's surface, owing to the diurnal rotation of the globe, are carried by the same alternately in and against the direction of translatory movement. From these results a great change in momentum which could be utilized in the simplest imaginable manner to furnish motive effort in any habitable region of the world. I cannot find words to describe my disappointment when later I realized that I was in the predicament of Archimedes, who vainly sought for a fixed point in the universe.

At the termination of my vacation, I was sent to the Polytechnic School in Gratz, Styria, which my father had chosen as one of the oldest and best reputed institutions. That was the moment I had eagerly awaited, and I began my studies under good auspices and firmly resolved to succeed. My previous training was above the average, due to my father's teaching and opportunities afforded. I had acquired the knowledge of a number of languages and waded thru the books of several libraries, picking up information more or less useful. Then again, for the first time, I could choose my subjects as I liked, and free-hand drawing was to bother me no more.

I had made up my mind to give my parents a surprise, and during the whole first year I regularly started my work at three o'clock in the morning and continued until eleven at night, no Sundays or holidays excepted. As most of my fellow-students took thinks easily, naturally enough I eclipsed all records. In the course of that year, I passed thru nine exams and the professors thought I deserved more than the highest qualifications. Armed with their flattering certificates, I went home for a short rest, expecting a triumph, and was mortified when my father made light of these hard-won honours. That almost killed my ambition; but later, after he had died, I was pained to find a package of letters which the professors had written him to the effect that unless he took me away from the Institution I would be killed thru overwork.

Thereafter I devoted myself chiefly to physics, mechanics and mathematical studies, spending the hours of leisure in the libraries. I had a veritable rnania for finishing whatever I began, which often got me into difficulties. On one occasion I started to read the works of Voltaire when I learned, to my dismay, that there were close on one hundred large volumes in small print which that monster had written while drinking seventy-two cups of black coffee per diem. It had to be done, but when I laid aside the last book, I was very glad, and said, "Never more!"

My first year's showing had won me the appreciation and friendship of several professors. Among these were Prof. Rogner, who was teaching arithmetical subjects and geometry; Prof. Poeschl, who held the chair of theoretical and experimental physics, and Dr. Alle, who taught integral calculus and specialized in differential equations. This scientist was the most brilliant lecturer to whom I ever listened. He took a special interest in my progress and would frequently remain for an hour or two in the lecture room, giving me problems to solve, in which I delighted. To him I explained a flying machine I had conceived, not an illusionary invention, but one based on sound, scientific principles, which has become realizable thru my turbine and will soon be given to the world. Both Professors Rogner and Poeschl were curious men. The former had peculiar ways of expressing himself and whenever he did so there was a riot, followed by a long and embarrassing pause.

Prof. Poeschl was a methodical and thoroly grounded German. He had enormous feet and hands like the paws of a bear, but all of his experiments were skilfully performed with lock-like precision and without a miss.

It was in the second year of my studies that we received a Gramme dynamo from Paris, having the horseshoe form of a laminated field magnet, and a wire-wound armature with a commutator. It was connected up and various effects of the currents were shown. While Prof. Poeschl was making demonstrations, running the machine as a motor, the brushes gave trouble, sparking badly, and I observed that it might be possible to operate a motor without these appliances. But he declared that it could not be done and did me the honour of delivering a lecture on the subject, at the conclusion of which he remarked: "Mr. Tesla may accomplish great things, but he certainly never will do this. It would be equivalent to converting a steadily pulling force, like that of gravity, into a rotary effort. It is a perpetual motion scheme, an impossible idea." But instinct is something which transcends knowledge. We have, undoubtedly, certain finer fibers that enable us to perceive truths when logical deduction, or any other wilful effort of the brain, is futile. For a time, I wavered, imprest by the professor's authority, but soon became convinced I was right and undertook the task with all the fire and boundless confidence of youth.

I started by first picturing in my mind a direct-current machine, running it and following the changing flow of the currents in the armature. Then I would imagine an alternator and investigate the processes taking place in a similar manner. Next, I would visualize systems comprising motors and generators and operate them in various ways. The images I saw were to me perfectly real and tangible. All my remaining term in Gratz was passed in intense but fruitless efforts of this kind, and I almost came to the conclusion that the problem was insolvable.

In 1880 I went to Prague, Bohemia, carrying out my father's wish to complete my education at the University there. It was in that city that I made a decided advance, which consisted in detaching the commutator from the machine and studying the phenomena in this new aspect, but still without result. In the year following there was a sudden change in my views of life. I realized that my parents had been making too great sacrifices on my account and resolved to relieve them of the burden. The wave of the American telephone had just reached the European continent and the system was to be installed in Budapest, Hungary. It appeared an ideal opportunity, all the more as a friend of our family was at the head of the enterprise. It was here that I suffered the complete breakdown of the nerves to which I have referred.

What I experienced during the period of that illness surpasses all belief. My sight and hearing were always extraordinary. I could clearly discern objects in the distance when others saw no trace of them. Several times in my boyhood I saved the houses of our neighbours from fire by hearing the faint crackling sounds which did not disturb their sleep and calling for help.

In 1899, when I was past forty and carrying on my experiments in Colorado, I could hear very distinctly thunderclaps at a distance of 550 miles. The limit of audition for my young assistants was scarcely more than 150 miles. My ear was thus over thirteen times more sensitive. Yet at that time I was, so to speak, stone deaf in comparison with the acuteness of my hearing while under the nervous strain. In Budapest I could hear the ticking of a watch with three rooms between me and the timepiece. A fly alighting on a table in the room would cause a dull thud in my ear. A carriage passing at a distance of a few miles fairly shook my whole body. The whistle of a locomotive twenty or thirty miles away made the bench or chair on which I sat vibrate so strongly that the pain was unbearable. The ground under my feet trembled continuously. I had to support my bed on rubber cushions to get any rest at all. The roaring noises from near and far often produced the effect of spoken words which would have frightened me had I not been able to resolve them into their accidental components. The sun's rays, when periodically intercepted, would cause blows of such force on my brain that they would stun me. I had to summon all my will power to pass under a bridge or other structure as I experienced a crushing pressure on the skull. In the dark I had the sense of a bat and could detect the presence of an object at a distance of twelve feet by a peculiar creepy sensation on the forehead. My pulse varied from a few to two hundred and sixty beats and all the tissues of the body quivered with twitching's and tremors which was perhaps the hardest to bear. A renowned physician who gave me daily large doses of Bromide of Potassium pronounced my malady unique and incurable.

It is my eternal regret that I was not under the observation of experts in physiology and psychology at that time. I clung desperately to life, but never expected to recover. Can anyone believe that so hopeless a physical wreck could ever be transformed into a man of astonishing strength and tenacity, able to work thirty-eight years almost without a day's interruption, and find himself still strong and fresh in body and mind? Such is my case. A powerful desire to live and to continue the work, and the assistance of a devoted friend and athlete accomplished the wonder. My health returned and with it the vigour of mind. In attacking the problem again, I almost regretted that the struggle was soon to end. I had so much energy to spare. When I undertook the task, it was not with a resolve such as men often make. With me it was sacred vow, a question of life and death. I knew that I would perish if I failed. Now I felt that the battle was won. Back in the deep recesses of the brain was the solution, but I could not yet give it outward expression. One afternoon, which is ever present in my recollection, I was enjoying a walk with my friend in the City Park and reciting poetry. At that age I knew entire books by heart, word for word. One of these was Goethe's "Faust." The sun was just setting and reminded me of the glorious passage:

"Sie ruckt und weicht, der Tag ist uberlebt, Dort eilt sie hin und fordert neues Leben. Oh, dass kein Flugel mich vom Boden hebt Ihr nach und immer nach zu streben!

Ein schoner Traum indessen sie entweicht,

Ach, zu des Geistes Flugeln wird so leicht

Kein korperlicher Flugel sich gesellen!"

[The glow retreats, done is the day of toil;

It yonder hastes, new fields of life exploring;

Ah, that no wing can lift me from the soil

Upon its track to follow, follow soaring!

A glorious dream! though now the glories fade.

Alas! the wings that lift the mind no aid

Of wings to lift the body can bequeath me.]

As I uttered these inspiring words the idea came like a flash of lightning and in an instant the truth was revealed. I drew with a stick on the sand the diagrams shown six years later in my address before the American Institute of Electrical Engineers, and my companion understood them perfectly. The images I saw were wonderfully sharp and clear and had the solidity of metal and stone, so much so that I told him: "See my motor here; watch me reverse it." I cannot begin to describe my emotions. Pygmalion seeing his statue come to life could not have been more deeply moved. A thousand secrets of nature which I might have stumbled upon accidentally I would have given for that one which I had wrested from her against all odds and at the peril of my existence.

# IV. The Discovery Of The Tesla Coil And Transformer

For a while I gave myself up entirely to the intense enjoyment of picturing machines and devising new forms. It was a mental state of happiness about as complete as I have ever known in life. Ideas came in an uninterrupted stream and the only difficulty I had was to hold them fast. The pieces of apparatus I conceived were to me absolutely real and tangible in every detail, even to the minute marks and signs of wear. I delighted in imagining the motors constantly running, for in this way they presented to mind's eye a more fascinating sight. When natural inclination develops into a passionate desire, one advances towards his goal in seven-league boots. In less than two months I evolved virtually all the types of motors and modifications of the system which are now identified with my name. It was, perhaps, providential that the necessities of existence commanded a temporary halt to this consuming activity of the mind. I came to Budapest prompted by a premature report concerning the telephone enterprise and, as irony of fate willed it, I had to accept a position as draftsman in the Central Telegraph Office of the Hungarian Government at a salary which I deem it my privilege not to disclose! Fortunately, I soon won the interest of the Inspector-in-Chief and was thereafter employed on calculations, designs and estimates in connection with new installations, until the Telephone Exchange was started, when I took charge of the same. The knowledge and practical experience I gained in the course of this work was most valuable and the employment gave me ample opportunities for the exercise of my inventive faculties. I made several improvements in the Central Station apparatus and perfected a telephone repeater or amplifier which was never patented or publicly described but would be creditable to me even today. In recognition of my efficient assistance the organizer of the undertaking, Mr. Puskas, upon disposing of his business in Budapest, offered me a position in Paris which I gladly accepted.

I never can forget the deep impression that magic city produced on my mind. For several days after my arrival, I roamed thru the streets in utter bewilderment of the new spectacle. The attractions were many and irresistible, but, alas, the income was spent as soon as received. When Mr. Puskas asked me how I was getting along in the new sphere, I described the situation accurately in the statement that "the last twenty-nine days of the month are the toughest!" I led a rather strenuous life in what would now be termed "Rooseveltian fashion." Every morning, regardless of weather, I would go from the Boulevard St. Marcel, where I resided, to a bathing house on the Seine, plunge into the water, loop the circuit twenty-seven times and then walk an hour to reach Ivry, where the Company's factory was located. There I would have a woodchopper's breakfast at half-past seven o'clock and then eagerly await the lunch hour, in the meanwhile cracking hard nuts for the Manager of the Works, Mr. Charles Batchellor, who was an intimate friend and assistant of Edison. Here I was thrown in contact with a few Americans who fairly fell in love with me because of my proficiency in billiards. To these men I explained my invention and one of them, Mr. D. Cunningham, Foreman of the Mechanical Department, offered to form a stock company. The proposal seemed to me comical in the extreme. I did not have the faintest conception of what that meant except that it was an American way of doing things. Nothing came of it, however, and during the next few months I had to travel from one to another place in France and Germany to cure the ills of the power plants. On my return to Paris, I submitted to one of the administrators of the Company, Mr. Rau, a plan for improving their dynamos and was given an opportunity. My success was complete, and the delighted directors accorded me the privilege of developing automatic regulators which were much desired. Shortly after there was some trouble with the lighting plant which had been installed at the new railroad station in Strassburg, Alsace. The wiring was defective and on the occasion of the opening ceremonies a large part of a wall was blown out thru a short-circuit right in the presence of old Emperor William I. The German Government refused to take the plant and the French Company was facing a serious loss. On account of my knowledge of the German language and past experience, I was entrusted with the difficult task of straightening out matters and early in 1883 I went to Strassburg on that mission.

Some of the incidents in that city have left an indelible record on my memory. By a curious coincidence, a number of men who subsequently achieved fame, lived there about that time. In later life I used to say, "There were bacteria of greatness in that old town. Others caught the disease, but I escaped!" The practical work, correspondence, and conferences with officials kept me preoccupied day and night, but, as soon as I was able to manage, I undertook the construction of a simple motor in a mechanical shop opposite the railroad station, having brought with me from Paris some material for that purpose. The consummation of the experiment was, however, delayed until the summer of that year when I finally had the satisfaction of seeing rotation effected by alternating currents of different phase, and without sliding contacts or commutator, as I had conceived a year before. It was an exquisite pleasure but not to compare with the delirium of joy following the first revelation.

Among my new friends was the former Mayor of the city, Mr. Bauzin, whom I had already in a measure acquainted with this and other inventions of mine and whose support I endeavoured to enlist. He was sincerely devoted to me and put my project before several wealthy persons but, to my mortification, found no response. He wanted to help me in every possible way and the approach of the first of July 1919, happens to remind me of a form of "assistance" I received from that charming man, which was not financial but none the less appreciated. In 1870, when the Germans invaded the country, Mr. Bauzin had buried a good-sized allotment of St. Estephe of 1801 and he came to the conclusion that he knew no worthier person than myself to consume that precious beverage. This, I may say, is one of the unforgettable incidents to which I have referred. My friend urged me to return to Paris as soon as possible and seek support there. This I was anxious to do but my work and negotiations were protracted owing to all sorts of petty obstacles I encountered so that at times the situation seemed hopeless.

Just to give an idea of German thoroness and "efficiency," I may mention here a rather funny experience. An incandescent lamp of 16 c.p. was to be placed in a hallway and upon selecting the proper location I ordered the monteur to run the wires. After working for a while, he concluded that the engineer had to be consulted and this was done. The latter made several objections but ultimately agreed that the lamp should be placed two inches from the spot I had assigned, whereupon the work proceeded. Then the engineer became worried and told me that Inspector Averdeck should be notified. That important person called, investigated, debated, and decided that the lamp should be shifted back two inches, which was the place I had marked. It was not long, however, before Averdeck got cold feet himself and advised me that he had informed Ober-Inspector Hieronimus of the matter and that I should await his decision. It was several days before the Ober-Inspector was able to free himself of other pressing duties but at last he arrived, and a two-hour debate followed, when he decided to move the lamp two inches farther. My hopes that this was the final act were shattered when the Ober-Inspector returned and said to me: "Regierungsrath Funke is so particular that I would not dare to give an order for placing this lamp without his explicit approval." Accordingly, arrangements for a visit from that great man were made. We started cleaning up and polishing early in the morning. Everybody brushed up, I put on my gloves and when Funke came with his retinue he was ceremoniously received. After two hours' deliberation he suddenly exclaimed: "I must be going," and pointing to a place on the ceiling, he ordered me to put the lamp there. It was the exact spot which I had originally chosen,

So, it went day after day with variations, but I was determined to achieve at whatever cost and in the end my efforts were rewarded. By the spring of 1884 all the differences were adjusted, the plant formally accepted, and I returned to Paris with pleasing anticipations. One of the administrators had promised me a liberal compensation in case I succeeded, as well as a fair consideration of the improvements I had made in their dynamos, and I hoped to realize a substantial sum. There were three administrators whom I shall designate as A, B and C for convenience. When I called on A, he told me that B had the say. This gentleman thought that only C could decide, and the latter was quite sure that A alone had the power to act. After several laps of this circulus vivios it dawned upon me that my reward was a castle in Spain. The utter failure of my attempts to raise capital for development was another disappointment and when Mr. Batchellor prest me to go to America with a view of redesigning the Edison machines, I determined to try my fortunes in the Land of Golden Promise. But the chance was nearly mist. I liquefied my modest assets, secured accommodations and found myself at the railroad station as the train was pulling out. At that moment I discovered that my money and tickets were gone. What to do was the question. Hercules had plenty of time to deliberate but I had to decide while running alongside the train with opposite feelings surging in my brain like condenser oscillations. Resolve, helped by dexterity, won out in the nick of time and upon passing thru the usual experiences, as trivial as unpleasant, I managed to embark for New York with the remnants of my belongings, some poems and articles I had written, and a package of calculations relating to solutions of an unsolvable integral and to my flying machine. During the voyage I sat most of the time at the stern of the ship watching for an opportunity to save somebody from a watery grave, without the slightest thought of danger. Later when I had absorbed some of the practical American sense I shivered at the recollection and marvelled at my former folly.

I wish that I could put in words my first impressions of this country. In the Arabian Tales I read how genii transported people into a land of dreams to live thru delightful adventures. My case was just the reverse. The genii had carried me from a world of dreams into one of realities. What I had left was beautiful, artistic and fascinating in every way; what I saw here was machined, rough and unattractive. A burly policeman was twirling his stick which looked to me as big as a log. I approached him politely with the request to direct me. "Six blocks down, then to the left," he said, with murder in his eyes. "Is this America?" I asked myself in painful surprise. "It is a century behind Europe in civilization." When I went abroad in 1889 - five years having elapsed since my arrival here I became convinced that it was more than one hundred years AHEAD of Europe and nothing has happened to this day to change my opinion.

The meeting with Edison was a memorable event in my life. I was amazed at this wonderful man who, without early advantages and scientific training, had accomplished so much. I had studied a dozen languages, delved in literature and art, and had spent my best years in libraries reading all sorts of stuff that fell into my hands, from Newton's "Principia" to the novels of Paul de Kock, and felt that most of my life had been squandered. But it did not take long before I recognized that it was the best thing I could have done. Within a few weeks I had won Edison's confidence and it came about in this way.

The S.S. Oregon, the fastest passenger steamer at that time, had both of its lighting machines disabled and its sailing was delayed. As the superstructure had been built after their installation it was impossible to remove them from the hold. The predicament was a serious one and Edison was much annoyed. In the evening I took the necessary instruments with me and went aboard the vessel where I stayed for the night. The dynamos were in bad condition, having several short-circuits and breaks, but with the assistance of the crew I succeeded in putting them in good shape. At five o'clock in the morning, when passing along Fifth Avenue on my way to the shop, I met Edison with Batchellor and a few others as they were returning home to retire. "Here is our Parisian running around at night," he said. When I told him that I was coming from the Oregon and had repaired both machines, he looked at me in silence and walked away without another word. But when he had gone some distance, I heard him remark: "Batchellor, this is a d-n good man," and from that time on I had full freedom in directing the work. For nearly a year my regular hours were from 10.30 A.M. until 5 o'clock the next morning without a day's exception. Edison said to me: "I have had many hard-working assistants, but you take the cake." During this period, I designed twenty-four different types of standard machines with short cores and of uniform pattern which replaced the old ones. The Manager had promised me fifty thousand dollars on the completion of this task, but it turned out to be a practical joke. This gave me a painful shock and I resigned my position.

Immediately thereafter some people approached me with the proposal of forming an arc light company under my name, to which I agreed. Here finally was an opportunity to develop the motor, but when I broached the subject to my new associates they said: "No, we want the arc lamp. We don't care for this alternating current of yours." In 1886 my system of arc lighting was perfected and adopted for factory and municipal lighting, and I was free, but with no other possession than a beautifully engraved certificate of stock of hypothetical value. Then followed a period of struggle in the new medium for which I was not fitted, but the reward came in the end and in April 1887, the Tesla Electric Company was organized, providing a laboratory and facilities. The motors I built there were exactly as I had imagined them. I made no attempt to improve the design, but merely reproduced the pictures as they appeared to my vision and the operation was always as I expected.

In the early part of 1888, an arrangement was made with the Westinghouse Company for the manufacture of the motors on a large scale. But great difficulties had still to be overcome. My system was based on the use of low frequency currents and the Westinghouse experts had adopted 133 cycles with the object of securing advantages in the transformation. They did not want to depart from their standard forms of apparatus and my efforts had to be concentrated upon adapting the motor to these conditions. Another necessity was to produce a motor capable of running efficiently at this frequency on two wires which was not easy of accomplishment.

At the close of 1889, however, my services in Pittsburg being no longer essential, I returned to New York and resumed experimental work in a laboratory on Grand Street, where I began immediately the design of high frequency machines. The problems of construction in this unexplored field were novel and quite peculiar and I encountered many difficulties. I rejected the inductor type, fearing that it might not yield perfect sine waves which were so important to resonant action. Had it not been for this I could have saved myself a great deal of labour. Another discouraging feature of the high frequency alternator seemed to be the inconstancy of speed which threatened to impose serious limitations to its use. I had already noted in my demonstrations before the American Institution of Electrical Engineers that several times the tune was lost, necessitating readjustment, and did not yet foresee, what I discovered long afterwards, a means of operating a machine of this kind at a speed constant to such a degree as not to vary more than a small fraction of one revolution between the extremes of load.

From many other considerations it appeared desirable to invent a simpler device for the production of electric oscillations. In 1856 Lord Kelvin had exposed the theory of the condenser discharge, but no practical application of that important knowledge was made. I saw the possibilities and undertook the development of induction apparatus on this principle. My progress was so rapid as to enable me to exhibit at my lecture in 1891 a coil giving sparks of five inches. On that occasion I frankly told the engineers of a defect involved in the transformation by the new method, namely, the loss in the spark gap. Subsequent investigation showed that no matter what medium is employed, be it air, hydrogen, mercury vapor, oil or a stream of electrons, the efficiency is the same. It is a law very much like that governing the conversion of mechanical energy. We may drop a weight from a certain height vertically down or carry it to the lower level along any devious path, it is immaterial insofar as the amount of work is concerned. Fortunately, however, this drawback is not fatal as by proper proportioning of the resonant circuits an efficiency of 85 per cent is attainable. Since my early announcement of the invention, it has come into universal use and wrought a revolution in many departments. But a still greater future awaits it. When in 1900 I obtained powerful discharges of 100 feet and flashed a current around the globe, I was reminded of the first tiny spark I observed in my Grand Street laboratory and was thrilled by sensations akin to those I felt when I discovered the rotating magnetic field.

# V. The Magnifying Transmitter

As I review the events of my past life, I realize how subtle are the influences that shape our destinies. An incident of my youth may serve to illustrate. One winter's day I managed to climb a steep mountain, in company with other boys. The snow was quite deep, and a warm southerly wind made it just suitable for our purpose. We amused ourselves by throwing balls which would roll down a certain distance, gathering more or less snow, and we tried to outdo one another in this exciting sport. Suddenly a ball was seen to go beyond the limit, swelling to enormous proportions until it became as big as a house and plunged thundering into the valley below with a force that made the ground tremble. I looked on spellbound, incapable of understanding what had happened. For weeks afterward the picture of the avalanche was before my eyes, and I wondered how anything so small could grow to such an immense size. Ever since that time the magnification of feeble actions fascinated me, and when, years later, I took up the experimental study of mechanical and electrical resonance, I was keenly interested from the very start. Possibly, had it not been for that early powerful impression, I might not have followed up the little spark I obtained with my coil and never developed my best invention, the true history of which I'll tell here for the first time.

"Lion hunters" have often asked me which of my discoveries I prize most. This depends on the point of view. Not a few technical men, very able in their special departments, but dominated by a pedantic spirit and near-sighted, have asserted that excepting the induction motor I have given to the world little of practical use. This is a grievous mistake. A new idea must not be judged by its immediate results. My alternating system of power transmission came at a psychological moment, as a long-sought answer to pressing industrial questions, and although considerable resistance had to be overcome and opposing interests reconciled, as usual, the commercial introduction could not be long delayed. Now, compare this situation with that confronting my turbine, for example. One should think that so simple and beautiful an invention, possessing many features of an ideal motor, should be adopted at once and, undoubtedly, it would under similar conditions. But the prospective effect of the rotating field was not to render worthless existing machinery; on the contrary, it was to give it additional value. The system lent itself to new enterprise as well as to improvement of the old. My turbine is an advance of a character entirely different. It is a radical departure in the sense that its success would mean the abandonment of the antiquated types of prime movers on which billions of dollars have been spent. Under such circumstances the progress must needs be slow and perhaps the greatest impediment is encountered in the prejudicial opinions created in the minds of experts by organized opposition.

Only the other day I had a disheartening experience when I met my friend and former assistant, Charles F. Scott, now professor of Electrical Engineering at Yale. I had not seen him for a long time and was glad to have an opportunity for a little chat at my office. Our conversation naturally enough drifted on my turbine, and I became heated to a high degree. "Scott," I exclaimed, carried away by the vision of a glorious future, "my turbine will scrap all the heat-engines in the world." Scott stroked his chin and looked away thoughtfully, as though making a mental calculation. "That will make quite a pile of scrap," he said, and left without another word!

These and other inventions of mine, however, were nothing more than steps forward in certain directions. In evolving them I simply followed the inborn sense to improve the present devices without any special thought of our far more imperative necessities. The "Magnifying Transmitter" was the product of labours extending through years, having for their chief object the solution of problems which are infinitely more important to mankind than mere industrial development.

If my memory serves me right, it was in November 1890, that I performed a laboratory experiment which was one of the most extraordinary and spectacular ever recorded in the annals of science. In investigating the behaviour of high frequency currents, I had satisfied myself that an electric field of sufficient intensity could be produced in a room to light up electrodeless vacuum tubes. Accordingly, a transformer was built to test the theory and the first trial proved a marvellous success. It is difficult to appreciate what those strange phenomena meant at that time. We crave for new sensations but soon become indifferent to them. The wonders of yesterday are today common occurrences. When my tubes were first publicly exhibited, they were viewed with amazement impossible to describe. From all parts of the world, I received urgent invitations and numerous honours and other flattering inducements were offered to me, which I declined.

But in 1892 the demands became irresistible, and I went to London where I delivered a lecture before the Institution of Electrical Engineers. It had been my intention to leave immediately for Paris in compliance with a similar obligation, but Sir James Dewar insisted on my appearing before the Royal Institution. I was a man of firm resolve but succumbed easily to the forceful arguments of the great Scotsman. He pushed me into a chair and poured out half a glass of a wonderful brown fluid which sparkled in all sorts of iridescent colours and tasted like nectar. "Now," said he. "You are sitting in Faraday's chair, and you are enjoying whiskey he used to drink." In both aspects it was an enviable experience. The next evening, I gave a demonstration before that Institution, at the termination of which Lord Rayleigh addressed the audience and his generous words gave me the first start in these endeavours. I fled from London and later from Paris to escape favours showered upon me and journeyed to my home where I passed through a most painful ordeal and illness. Upon regaining my health, I began to formulate plans for the resumption of work in America. Up to that time I never realized that I possessed any particular gift of discovery but Lord Rayleigh, whom I always considered as an ideal man of science, had said so and if that was the case, I felt that I should concentrate on some big idea.

One day, as I was roaming in the mountains, I sought shelter from an approaching storm. The sky became overhung with heavy clouds but somehow the rain was delayed until, all of a sudden, there was a lightning flash and a few moments after a deluge. This observation set me thinking. It was manifest that the two phenomena were closely related, as cause and effect, and a little reflection led me to the conclusion that the electrical energy involved in the precipitation of the water was inconsiderable, the function of lightning being much like that of a sensitive trigger.

Here was a stupendous possibility of achievement. If we could produce electric effects of the required quality, this whole planet and the conditions of existence on it could be transformed. The sun raises the water of the oceans and winds drive it to distant regions where it remains in a state of most delicate balance. If it were in our power to upset it when and wherever desired, this mighty life-sustaining stream could be at will controlled. We could irrigate arid deserts, create lakes and rivers and provide motive power in unlimited amounts. This would be the most efficient way of harnessing the sun to the uses of man. The consummation depended on our ability to develop electric forces of the order of those in nature. It seemed a hopeless undertaking, but I made up my mind to try it and immediately on my return to the United States, in the Summer of 1892, work was begun which was to me all the more attractive, because a means of the same kind was necessary for the successful transmission of energy without wires.

The first gratifying result was obtained in the spring of the succeeding year when I reached tensions of about 1,000,000 volts with my conical coil. That was not much in the light of the present art, but it was then considered a feat. Steady progress was made until the destruction of my laboratory by fire in 1895, as may be judged from an article by T. C. Martin which appeared in the April number of the Century Magazine. This calamity set me back in many ways and most of that year had to be devoted to planning and reconstruction. However, as soon as circumstances permitted, I returned to the task.

Although I knew that higher electro-motive forces were attainable with apparatus of larger dimensions, I had an instinctive perception that the object could be accomplished by the proper design of a comparatively small and compact transformer. In carrying on tests with a secondary in the form of a flat spiral, as illustrated in my patents, the absence of streamers surprised me, and it was not long before I discovered that this was due to the position of the turns and their mutual action. Profiting from this observation I resorted to the use of a high-tension conductor with turns of considerable diameter sufficiently separated to keep down the distributed capacity, while at the same time preventing undue accumulation of the charge at any point. The application of this principle enabled me to produce pressures of 4,000,000 volts, which was about the limit obtainable in my new laboratory at Houston Street, as the discharges extended through a distance of 16 feet. A photograph of this transmitter was published in the Electrical Review of November 1898.

In order to advance further along this line, I had to go into the open, and in the spring of 1899, having completed preparations for the erection of a wireless plant, I went to Colorado where I remained for more than one year. Here I introduced other improvements and refinements which made it possible to generate currents of any tension that may be desired. Those who are interested will find some information in regard to the experiments I conducted there in my article, "The Problem of Increasing Human Energy" in the Century Magazine of

June 1900, to which I have referred on a previous occasion.

I have been asked by the ELECTRICAL EXPERIMENTER to be quite explicit on this subject so that my young friends among the readers of the magazine will clearly understand the construction and operation of my "Magnifying Transmitter" and the purposes for which it is intended. Well, then, in the first place, it is a resonant transformer with a secondary in which the parts, charged to a high potential, are of considerable area and arranged in space along ideal enveloping surfaces of very large radii of curvature, and at proper distances from one another thereby insuring a small electric surface density everywhere so that no leak can occur even if the conductor is bare. It is suitable for any frequency, from a few to many thousands of cycles per second, and can be used in the production of currents of tremendous volume and moderate pressure, or of smaller amperage and immense electromotive force. The maximum electric tension is merely dependent on the curvature of the surfaces on which the charged elements are situated and the area of the latter.

Judging from my past experience, as much as 100,000,000 volts are perfectly practicable. On the other hand, currents of many thousands of amperes may be obtained in the antenna. A plant of but very moderate dimensions is required for such performances.

Theoretically, a terminal of less than 90 feet in diameter is sufficient to develop an electromotive force of that magnitude while for antenna currents of from 2,000-4,000 amperes at the usual frequencies it need not be larger than 30 feet in diameter.

In a more restricted meaning this wireless transmitter is one in which the Hertz-wave radiation is an entirely negligible quantity as compared with the whole energy, under which condition the damping factor is extremely small and an enormous charge is stored in the elevated capacity. Such a circuit may then be excited with impulses of any kind, even of low frequency and it will yield sinusoidal and continuous oscillations like those of an alternator.

Taken in the narrowest significance of the term, however, it is a resonant transformer which, besides possessing these qualities, is accurately proportioned to fit the globe and its electrical constants and properties, by virtue of which design it becomes highly efficient and effective in the wireless transmission of energy. Distance is then absolutely eliminated, there being no diminution in the intensity of the transmitted impulses. It is even possible to make the actions increase with the distance from the plant according to an exact mathematical law.

This invention was one of a number comprised in my "World-System" of wireless transmission which I undertook to commercialize on my return to New York in 1900. As to the immediate purposes of my enterprise, they were clearly outlined in a technical statement of that period from which I quote:

"The 'World-System' has resulted from a combination of several original discoveries made by the inventor in the course of long continued research and experimentation. It makes possible not only the instantaneous and precise wireless transmission of any kind of signals, messages or characters, to all parts of the world, but also the inter-connection of the existing telegraph, telephone, and other signal stations without any change in their present equipment. By its means, for instance, a telephone subscriber here may call up and talk to any other subscriber on the Globe. An inexpensive receiver, not bigger than a watch, will enable him to listen anywhere, on land or sea, to a speech delivered or music played in some other place, however distant. These examples are cited merely to give an idea of the possibilities of this great scientific advance, which annihilates distance and makes that perfect natural conductor, the Earth, available for all the innumerable purposes which human ingenuity has found for a line-wire. One far-reaching result of this is that any device capable of being operated thru one or more wires (at a distance obviously restricted) can likewise be actuated, without artificial conductors and with the same facility and accuracy, at distances to which there are no limits other than those imposed by the physical dimensions of the Globe. Thus, not only will entirely new fields for commercial exploitation be opened up by this ideal method of transmission but the old ones vastly extended.

The 'World-System' is based on the application of the following important inventions and discoveries:

1. The 'Tesla Transformer.' This apparatus is in the production of electrical vibrations as revolutionary as gunpowder was in warfare. Currents many times stronger than any ever generated in the usual ways, and sparks over one hundred feet long, have been produced by the inventor with an instrument of this kind.

2. The 'Magnifying Transmitter.' This is Tesla's best invention, a peculiar transformer specially adapted to excite the Earth, which is in the transmission of electrical energy what the telescope is in astronomical observation. By the use of this marvellous device, he has already set up electrical movements of greater intensity than those of lightning and passed a current, sufficient to light more than two hundred incandescent lamps, around the Globe.

3. The 'Tesla Wireless System.' This system comprises a number of improvements and is the only means known for transmitting economically electrical energy to a distance without wires. Careful tests and measurements in connection with an experimental station of great activity, erected by the inventor in Colorado, have demonstrated that power in any desired amount can be conveyed, clear across the Globe, if necessary, with a loss not exceeding a few per cent.

4. The 'Art of Individualization.' This invention of Tesla's is to primitive 'tuning' what refined language is to unarticulated expression. It makes possible the transmission of signals or messages absolutely secret and exclusive both in the active and passive aspect, that is, non-interfering as well as non-interferable. Each signal is like an individual of unmistakable identity and there is virtually no limit to the number of stations or instruments which can be simultaneously operated without

the slightest mutual disturbance.

5. 'The Terrestrial Stationary Waves.' This wonderful discovery, popularly explained, means that the Earth is responsive to electrical vibrations of definite pitch just as a tuning fork to certain waves of sound. These particular electrical vibrations, capable of powerfully exciting the Globe, lend themselves to innumerable uses of great importance commercially and in many other respects.

The first 'World-System' power plant can be put in operation in nine months. With this power plant it will be practicable to attain electrical activities up to ten million horsepower and it is designed to serve for as many technical achievements as are possible without due expense. Among these the following may be mentioned:

(1) The inter-connection of the existing telegraph exchanges or offices all over the world.

(2) The establishment of a secret and non-interferable government telegraph service.

(3) The inter-connection of all the present telephone exchanges or offices on the Globe.

(4) The universal distribution of general news, by telegraph or telephone, in connection with the Press.

(5) The establishment of such a 'World-System' of intelligence transmission for exclusive private use.

(6) The inter-connection and operation of all stock tickers of the world.

(7) The establishment of a 'World-System' of musical distribution, etc.

(8) The universal registration of time by cheap clocks indicating the hour with astronomical precision and requiring no attention whatever.

(9) The world transmission of typed or handwritten characters,

letters, checks, etc.

(10) The establishment of a universal marine service enabling the navigators of all ships to steer perfectly without compass, to determine the exact location, hour and speed, to prevent collisions and disasters, etc.

(11) The inauguration of a system of world-printing on land and sea.

(12) The world reproduction of photographic pictures and all kinds of drawings or records."

I also proposed to make demonstrations in the wireless transmission of power on a small scale but sufficient to carry conviction. Besides these I referred to other and incomparably more important applications of my discoveries which will be disclosed at some future date.

A plant was built on Long Island with a tower 187 feet high, having a spherical terminal about 68 feet in diameter. These dimensions were adequate for the transmission of virtually any amount of energy. Originally only from 200 to 300 K.W. were provided but I intended to employ later several thousand horsepower. The transmitter was to emit a wave complex of special characteristics and I had devised a unique method of telephonic control of any amount of energy.

The tower was destroyed two years ago but my projects are being developed and another one, improved in some features, will be constructed. On this occasion I would contradict the widely circulated report that the structure was demolished by the Government which owing to war conditions, might have created prejudice in the minds of those who may not know that the papers, which thirty years ago conferred upon me the honour of American citizenship, are always kept in a safe, while my orders, diplomas, degrees, gold medals and other distinctions are packed away in old trunks. If this report had a foundation, I would have been refunded a large sum of money which I expended in the construction of the tower. On the contrary it was in the interest of the Government to preserve it, particularly as it would have made possible to mention just one valuable result, the location of a submarine in any part of the world. My plant, services, and all my improvements have always been at the disposal of the officials and ever since the outbreak of the European conflict I have been working at a sacrifice on several inventions of mine relating to aerial navigation, ship propulsion and wireless transmission which are of the greatest importance to the country. Those who are well informed know that my ideas have revolutionized the industries of the United States and I am not aware that there lives an inventor who has been, in this respect, as fortunate as myself especially as regards the use of his improvements in the war. I have refrained from publicly expressing myself on this subject before as it seemed improper to dwell on personal matters while all the world was in dire trouble.

I would add further, in view of various rumours which have reached me, that Mr. J. Pierpont Morgan did not interest himself with me in a business way but in the same large spirit in which he has assisted many other pioneers. He carried out his generous promise to the letter and it would have been most unreasonable to expect from him anything more. He had the highest regard for my attainments and gave me every evidence of his complete faith in my ability to ultimately achieve what I had set out to do. I am unwilling to accord to some smallminded and jealous individuals the satisfaction of having thwarted my efforts. These men are to me nothing more than microbes of a nasty disease. My project was retarded by laws of nature. The world was not prepared for it. It was too far ahead of time. But the same laws will prevail in the end and make it a triumphal success.

# VI. The Art Of Telautomatics

No subject to which I have ever devoted myself has called for such concentration of mind and strained to so dangerous a degree the finest fibres of my brain as the system of which the Magnifying Transmitter is the foundation. I put all the intensity and vigour of youth in the development of the rotating field discoveries, but those early labours were of a different character. Although strenuous in the extreme, they did not involve that keen and exhausting discernment which had to be exercised in attacking the many puzzling problems of the wireless. Despite my rare physical endurance at that period the abused nerves finally rebelled, and I suffered a complete collapse, just as the consummation of the long and difficult task was almost in sight.

Without doubt I would have paid a greater penalty later, and very likely my career would have been prematurely terminated, had not providence equipped me with a safety device, which has seemed to improve with advancing years and unfailingly comes into play when my forces are at an end. So long as it operates, I am safe from danger, due to overwork, which threatens other inventors and, incidentally, I need no vacations which are indispensable to most people. When I am all but used up, I simply do as the darkies, who "naturally fall asleep while white folks worry." To venture a theory out of my sphere, the body probably accumulates little by little a definite quantity of some toxic agent and I sink into a nearly lethargic state which lasts half an hour to the minute. Upon awakening I have the sensation as though the events immediately preceding had occurred very long ago, and if I attempt to continue the interrupted train of thought I feel a veritable mental nausea. Involuntarily I then turn to other work and am surprised at the freshness of the mind and ease with which I overcome obstacles that had baffled me before. After weeks or months my passion for the temporarily abandoned invention returns and I invariably find answers to all the vexing questions with scarcely any effort.

In this connection I will tell of an extraordinary experience which may be of interest to students of psychology. I had produced a striking phenomenon with my grounded transmitter and was endeavouring to ascertain its true significance in relation to the currents propagated through the earth. It seemed a hopeless undertaking, and for more than a year I worked unremittingly, but in vain. This profound study so entirely absorbed me that I became forgetful of everything else, even of my undermined health. At last, as I was at the point of breaking down, nature applied the preservative inducing lethal sleep. Regaining my senses, I realized with consternation that I was unable to visualize scenes from my life except those of infancy, the very first ones that had entered my consciousness. Curiously enough, these appeared before my vision with startling distinctness and afforded me welcome relief. Night after night, when retiring, I would think of them and more and more of my previous existence was revealed. The image of my mother was always the principal figure in the spectacle that slowly unfolded, and a consuming desire to see her again gradually took possession of me. This feeling grew so strong that I resolved to drop all work and satisfy my longing. But I found it too hard to break away from the laboratory, and several months elapsed during which I had succeeded in reviving all the impressions of my past life up to the spring of 1892. In the next picture that came out of the mist of oblivion, I saw myself at the Hotel de la Paix in Paris just coming to from one of my peculiar sleeping spells, which had been caused by prolonged exertion of the brain. Imagine the pain and distress I felt when it flashed upon my mind that a dispatch was handed to me at that very moment bearing the sad news that my mother was dying. I remembered how I made the long journey home without an hour of rest and how she passed away after weeks of agony! It was especially remarkable that during all this period of partially obliterated memory I was fully alive to everything touching on the subject of my research. I could recall the smallest details and the least significant observations in my experiments and even recite pages of text and complex mathematical formulae.

My belief is firm in a law of compensation. The true rewards are ever in proportion to the labour and sacrifices made. This is one of the reasons why I feel certain that of all my inventions, the Magnifying Transmitter will prove most important and valuable to future generations. I am prompted to this prediction not so much by thoughts of the commercial and industrial revolution which it will surely bring about, but of the humanitarian consequences of the many achievements it makes possible. Considerations of mere utility weigh little in the balance against the higher benefits of civilization. We are confronted with portentous problems which cannot be solved just by providing for our material existence, however abundantly. On the contrary, progress in this direction is fraught with hazards and perils not less menacing than those born from want and suffering. If we were to release the energy of atoms or discover some other way of developing cheap and unlimited power at any point of the globe this accomplishment, instead of being a blessing, might bring disaster to mankind in giving rise to dissension and anarchy which would ultimately result in the enthronement of the hated regime of force. The greatest good will comes from technical improvements tending to unification and harmony, and my wireless transmitter is pre-eminently such. By it means the human voice and likeness will be reproduced everywhere and factories driven thousands of miles from waterfalls furnishing the power; aerial machines will be propelled around the earth without a stop and the sun's energy controlled to create lakes and rivers for motive purposes and transformation of arid deserts into fertile land. Its introduction for telegraphic, telephonic and similar uses will automatically cut out the statics and all other interferences which at present impose narrow limits to the application of the wireless.

This is a timely topic on which a few words might not be amiss. During the past decade a number of people have arrogantly claimed that they had succeeded in doing away with this impediment. I have carefully examined all of the arrangements described and tested most of them long before they were publicly disclosed, but the finding was uniformly negative. A recent official statement from the U.S. Navy may, perhaps, have taught some beguilable news editors how to appraise these announcements at their real worth. As a rule, the attempts are based on theories so fallacious that whenever they come to my notice I cannot help thinking in a lighter vein. Quite recently a new discovery was heralded, with a deafening flourish of trumpets, but it proved another case of a mountain bringing forth a mouse.

This reminds me of an exciting incident which took place years ago when I was conducting my experiments with currents of high frequency. Steve Brodie had just jumped off the Brooklyn Bridge. The feat has been vulgarized since by imitators, but the first report electrified New York. I was very impressionable then and frequently spoke of the daring printer. On a hot afternoon I felt the necessity of refreshing myself and stepped into one of the popular thirty thousand institutions of this great city where a delicious twelve per cent beverage was served which can now be had only by making a trip to the poor and devastated countries of Europe. The attendance was large and not over distinguished, and a matter was discussed which gave me an admirable opening for the careless remark: "This is what I said when I jumped off the bridge." No sooner had I uttered these words than I felt like the companion of Timotheus in the poem of Schiller. In an instant there was a pandemonium and a dozen voices cried: "It is Brodie! " I threw a quarter on the counter and bolted for the door but the crowd was at my heels with yells: "Stop, Steve!" which must have been misunderstood for many persons tried to hold me up as I ran frantically for my haven of refuge. By darting around corners, I fortunately managed - through the medium of a fire-escape - to reach the laboratory where I threw off my coat, camouflaged myself as a hard-working blacksmith, and started the forge. But these precautions proved unnecessary; I had eluded my pursuers. For many years afterward, at night, when imagination turns into spectres the trifling troubles of the day, I often thought, as I tossed on the bed, what my fate would have been had that mob caught me and found out that I was not Steve Brodie!

Now the engineer, who lately gave an account before a technical body of a novel remedy against statics based on a "heretofore unknown law of nature," seems to have been as reckless as myself when he contended that these disturbances propagate up and down, while those of a transmitter proceed along the earth. It would mean that a condenser, as this globe, with its gaseous envelope, could be charged and discharged in a manner quite contrary to the fundamental teachings propounded in every elemental textbook of physics. Such a supposition would have been condemned as erroneous, even in Franklin's time, for the facts bearing on this were then well known and the identity between atmospheric electricity and that developed by machines was fully established. Obviously, natural and artificial disturbances propagate through the earth and the air in exactly the same way, and both set up electromotive forces in the horizontal, as well as vertical, sense. Interference cannot be overcome by any such methods as were proposed. The truth is this: in the air the potential increases at the rate of about fifty volts per foot of elevation, owing to which there may be a difference of pressure amounting to twenty, or even forty thousand volts between the upper and lower ends of the antenna. The masses of the charged atmosphere are constantly in motion and give up electricity to the conductor, not continuously but rather disruptively, this producing a grinding noise in a sensitive telephonic receiver. The higher the terminal and the greater the space encompassed by the wires, the more pronounced is the effect, but it must be understood that it is purely local and has little to do with the real trouble.

In 1900, while perfecting my wireless system, one form of apparatus comprised four antennae. These were carefully calibrated to the same frequency and connected in multiple with the object of magnifying the action, in receiving from any direction. When I desired to ascertain the origin of the transmitted impulses, each diagonally situated pair was put in series with a primary coil energizing the detector circuit. In the former case the sound was loud in the telephone; in the latter it ceased, as expected, the two antennae neutralizing each other, but the true statics manifested themselves in both instances and I had to devise special preventives embodying different principles.

By employing receivers connected to two points of the ground, as suggested by me long ago, this trouble caused by the charged air, which is very serious in the structures as now built, is nullified and besides, the liability of all kinds of interference is reduced to about one-half, because of the directional character of the circuit. This was perfectly self-evident but came as a revelation to some simple-minded wireless folks whose experience was confined to forms of apparatus that could have been improved with an axe, and they have been disposing of the bear's skin before killing him. If it were true that strays performed such antics, it would be easy to get rid of them by receiving without aerials. But, as a matter of fact, a wire buried in the ground which, conforming to this view, should be absolutely immune, is more susceptible to certain extraneous impulses than one placed vertically in the air. To state it fairly, a slight progress has been made, but not by virtue of any particular method or device. It was achieved simply by discarding the enormous structures, which are bad enough for transmission but wholly unsuitable for reception and adopting a more appropriate type of receiver. As I pointed out in a previous article, to dispose of this difficulty for good, a radical change must be made in the system, and the sooner this is done the better.

It would be calamitous, indeed, if at this time when the art is in its infancy and the vast majority, not excepting even experts, have no conception of its ultimate possibilities, a measure would be rushed through the legislature making it a government monopoly. This was proposed a few weeks ago by Secretary Daniels, and no doubt that distinguished official has made his appeal to the Senate and House of Representatives with sincere conviction. But universal evidence unmistakably shows that the best results are always obtained in healthful commercial competition. There are, however, exceptional reasons why wireless should be given the fullest freedom of development. In the first place it offers prospects immeasurably greater and more vital to betterment of human life than any other invention or discovery in the history of man. Then again, it must be understood that this wonderful art has been, in its entirety, evolved here and can be called "American" with more right and propriety than the telephone, the incandescent lamp or the aeroplane. Enterprising press agents and stock jobbers have been so successful in spreading misinformation that even so excellent a periodical as the Scientific American accords the chief credit to a foreign country. The Germans, of course, gave us the Hertz-waves and the Russian, English, French and Italian experts were quick in using them for signalling purposes. It was an obvious application of the new agent and accomplished with the old classical and unimproved induction coil - scarcely anything more than another kind of heliography. The radius of transmission was very limited, the results attained of little value, and the Hertz oscillations, as a means for conveying intelligence, could have been advantageously replaced by soundwaves, which I advocated in 1891. Moreover, all of these attempts were made three years after the basic principles of the wireless system, which is universally employed to-day, and its potent instrumentalities had been clearly described and developed in America. No trace of those Hertzian appliances and methods remains today. We have proceeded in the very opposite direction and what has been done is the product of the brains and efforts of citizens of this country. The fundamental patents have expired, and the opportunities are open to all. The chief argument of the Secretary is based on interference. According to his statement, reported in the New York Herald of July 29th, signals from a powerful station can be intercepted in every village of the world. In view of this fact, which was demonstrated in my experiments of 1900, it would be of little use to impose restrictions in the United States.

As throwing light on this point, I may mention that only recently an odd-looking gentleman called on me with the object of enlisting my services in the construction of world transmitters in some distant land. "We have no money," he said, "but carloads of solid gold and we will give you a liberal amount." I told him that I wanted to see first what will be done with my inventions in America, and this ended the interview. But I am satisfied that some dark forces are at work, and as time goes on the maintenance of continuous communication will be rendered more difficult. The only remedy is a system immune against interruption. It has been perfected, it exists, and all that is necessary is to put it in operation.

The terrible conflict is still uppermost in the minds and perhaps the greatest importance will be attached to the Magnifying Transmitter as a machine for attack and defence, more particularly in connection with Telautomatics. This invention is a logical outcome of observations begun in my boyhood and continued throughout my life. When the first results were publish the Electrical Review stated editorially that it would become one of the "most potent factors in the advance and civilization of mankind." The time is not distant when this prediction will be fulfilled. In 1898 and 1900 it was offered to the Government and might have been adopted were I one of those who would go to Alexander's shepherd when they want a favour from Alexander. At that time, I really thought that it would abolish war, because of its unlimited destructiveness and exclusion of the personal element of combat. But while I have not lost faith in its potentialities, my views have changed since.

War cannot be avoided until the physical cause for its recurrence is removed and this, in the last analysis, is the vast extent of the planet on which we live. Only thru annihilation of distance in every respect, as the conveyance of intelligence, transport of passengers and supplies and transmission of energy will conditions be brought about someday, insuring permanency of friendly relations. What we now want most is closer contact and better understanding between individuals and communities all over the earth, and the elimination of that fanatic devotion to exalted ideals of national egoism and pride which is always prone to plunge the world into primeval barbarism and strife. No league or parliamentary act of any kind will ever prevent such a calamity. These are only new devices for putting the weak at the mercy of the strong. I have express myself in this regard fourteen years ago, when a combination of a few leading governments - a sort of Holy Alliance - was advocated by the late Andrew Carnegie, who may be fairly considered as the father of this idea, having given to it more publicity and impetus than anybody else prior to the efforts of the President. While it cannot be denied that such a pact might be of material advantage to some less fortunate peoples, it cannot attain the chief object sought. Peace can only come as a natural consequence of universal enlightenment and merging of races, and we are still far from this blissful realization.

As I view the world of today, in the light of the gigantic struggle we have witnessed, I am filled with conviction that the interests of humanity would be best served if the United States remained true to its traditions and kept out of "entangling alliances." Situated as it is, geographically, remote from the theatres of impending conflicts, without incentive to territorial aggrandizement, with inexhaustible resources and immense population thoroly imbued with the spirit of liberty and right, this country is placed in a unique and privileged position. It is thus able to exert, independently, its colossal strength and moral force to the benefit of all, more judiciously and effectively, than as member of a league.

In one of these biographical sketches, published in the ELECTRICAL EXPERIMENTER, I have dwelt on the circumstances of my early life and told of an affliction which compelled me to unremitting exercise of imagination and self-observation. This mental activity, at first involuntary under the pressure of illness and suffering, gradually became second nature and led me finally to recognize that I was but an automaton devoid of free will in thought and action and merely responsive to the forces of the environment. Our bodies are of such complexity of structure, the motions we perform are so numerous and involved, and the external impressions on our sense organs to such a degree delicate and elusive that it is hard for the average person to grasp this fact. And yet nothing is more convincing to the trained investigator than the mechanistic theory of life which had been, in a measure, understood and propounded by Descartes three hundred years ago. But in his time many important functions of our organism were unknown and, especially with respect to the nature of light and the construction and operation of the eye, philosophers were in the dark.

In recent years the progress of scientific research in these fields has been such as to leave no room for a doubt in regard to this view on which many works have been published. One of its ablest and most eloquent exponents is, perhaps, Felix Le Dantec, formerly assistant of Pasteur. Prof. Jacques Loeb has performed remarkable experiments in heliotropism, clearly establishing the controlling power of light in lower forms of organisms, and his latest book, "Forced Movements," is revelatory. But while men of science accept this theory simply as any other that is recognized, to me it is a truth which I hourly demonstrate by every act and thought of mine. The consciousness of the external impression prompting me to any kind of exertion, physical or mental, is ever present in my mind. Only on very rare occasions, when I was in a state of exceptional concentration, have I found difficulty in locating the original impulses.

The by far greater number of human beings are never aware of what is passing around and within them, and millions fall victims of disease and die prematurely just on this account. The commonest every-day occurrences appear to them mysterious and inexplicable. One may feel a sudden wave of sadness and rake his brain for an explanation when he might have noticed that it was caused by a cloud cutting off the rays of the sun. He may see the image of a friend dear to him under conditions which he construes as very peculiar, when only shortly before he has passed him in the street or seen his photograph somewhere. When he loses a collar button he fusses and swears for an hour, being unable to visualize his previous actions and locate the object directly. Deficient observation is merely a form of ignorance and responsible for the many morbid notions and foolish ideas prevailing. There is not more than one out of every ten persons who does not believe in telepathy and other psychic manifestations, spiritualism and communion with the dead, and who would refuse to listen to willing or unwilling deceivers.

Just to illustrate how deeply rooted this tendency has become even among the clearheaded American population, I may mention a comical incident. Shortly before the war, when the exhibition of my turbines in this city elicited widespread comment in the technical papers, I anticipated that there would. be a scramble among manufacturers to get hold of the invention, and I had particular designs on that man from Detroit who has an uncanny faculty for accumulating millions. So confident was I that he would turn up some day, that I declared this as certain to my secretary and assistants. Sure enough, one fine morning a body of engineers from the Ford Motor Company presented themselves with the request of discussing with me an important project. "Didn't I tell you?" I remarked triumphantly to my employees, and one of them said, "You are amazing, Mr. Tesla; everything comes out exactly as you predict." As soon as these hard-headed men were seated I, of course, immediately began to extol the wonderful features of my turbine, when the spokesmen interrupted me and said, "We know all about this, but we are on a special errand. We have formed a psychological society for the investigation of psychic phenomena, and we want you to join us in this undertaking." I suppose those engineers never knew how near they came to being fired out of my office.

Ever since I was told by some of the greatest men of the time, leaders in science whose names are immortal, that I am possessed of an unusual mind, I bent all my thinking faculties on the solution of great problems regardless of sacrifice. For many years I endeavoured to solve the enigma of death and watched eagerly for every kind of spiritual indication. But only once in the course of my existence have, I had an experience which momentarily impressed me as supernatural. It was at the time of my mother's death. I had become completely exhausted by pain and long vigilance, and one night was carried to a building about two blocks from our home. As I lay helpless there, I thought that if my mother died while I was away from her bedside, she would surely give me a sign. Two or three months before I was in London in company with my late friend, Sir William Crookes, when spiritualism was discussed, and I was under the full sway of these thoughts. I might not have paid attention to other men but was susceptible to his arguments as it was his epochal work on radiant matter, which I had read as a student, that made me embrace the electrical career. I reflected that the conditions for a look into the beyond were most favourable, for my mother was a woman of genius and particularly excelling in the powers of intuition. During the whole night every fibre in my brain was strained in expectancy, but nothing happened until early in the morning, when I fell in a sleep, or perhaps a swoon, and saw a cloud carrying angelic figures of marvellous beauty, one of whom gazed upon me lovingly and gradually assumed the features of my mother. The appearance slowly floated across the room and vanished, and I was awakened by an indescribably sweet song of many voices. In that instant a certitude, which no words can express, came upon me that my mother had just died. And that was true. I was unable to understand the tremendous weight of the painful knowledge I received in advance and wrote a letter to Sir William Crookes while still under the domination of these impressions and in poor bodily health. When I recovered, I sought for a long time the external cause of this strange manifestation and, to my great relief, I succeeded after many months of fruitless effort. I had seen the painting of a celebrated artist, representing allegorically one of the seasons in the form of a cloud with a group of angels which seemed to actually float in the air, and this had struck me forcefully. It was exactly the same that appeared in my dream, with the exception of my mother's likeness. The music came from the choir in the church nearby at the early mass of Easter morning, explaining everything satisfactorily in conformity with scientific facts.

This occurred long ago, and I have never had the faintest reason since to change my views on psychical and spiritual phenomena, for which there is absolutely no foundation. The belief in these is the natural outgrowth of intellectual development. Religious dogmas are no longer accepted in their orthodox meaning, but every individual cling to faith in a supreme power of some kind. We all must have an ideal to govern our conduct and ensure contentment, but it is immaterial whether it be one of creed, art, science or anything else, so long as it fulfils the function of a dematerializing force. It is essential to the peaceful existence of humanity as a whole that one common conception should prevail.

While I have failed to obtain any evidence in support of the contentions of psychologists and spiritualists, I have proved to my complete satisfaction the automatism of life, not only through continuous observations of individual actions, but even more conclusively through certain generalizations. These amount to a discovery which I consider of the greatest moment to human society, and on which I shall briefly dwell. I got the first inkling of this astounding truth when I was still a very young man, but for many years I interpreted what I noted simply as coincidences. Namely, whenever either myself or a person to whom I was attached, or a cause to which I was devoted, was hurt by others in a particular way, which might be best popularly characterized as the most unfair imaginable, I experienced a singular and undefinable pain which, for want of a better term, I have qualified as "cosmic," and shortly thereafter, and invariably, those who had inflicted it came to grief. After many such cases I confided this to a number of friends, who had the opportunity to convince themselves of the truth of the theory which I have gradually formulated, and which may be stated in the following few words:

Our bodies are of similar construction and exposed to the same external influences. This results in likeness of response and concordance of the general activities on which all our social and other rules and laws are based. We are automata entirely controlled by the forces of the medium being tossed about like corks on the surface of the water but mistaking the resultant of the impulses from the outside for free will. The movements and other actions we perform are always life preservative and tho seemingly quite independent from one another, we are connected by invisible links. So long as the organism is in perfect order it responds accurately to the agents that prompt it, but the moment that there is some derangement in any individual, his self-preservative power is impaired. Everybody understands, of course, that if one becomes deaf, has his eyesight weakened, or his limbs injured, the chances for his continued existence are lessened. But this is also true, and perhaps more so, of certain defects in the brain which deprive the automaton, more or less, of that vital quality and cause it to rush into destruction. A very sensitive and observant being, with his highly developed mechanism all intact, and acting with precision in obedience to the changing conditions of the environment, is endowed with a transcending mechanical sense, enabling him to evade perils too subtle to be directly perceived. When he comes in contact with others whose controlling organs are radically faulty, that sense asserts itself and he feels the "cosmic" pain. The truth of this has been borne out in hundreds of instances and I am inviting other students of nature to devote attention to this subject, believing that thru combined and systematic effort results of incalculable value to the world will be attained.

The idea of constructing an automaton, to bear out my theory, presented itself to me early but I did not begin active work until 1893, when I started my wireless investigations. During the succeeding two or three years a number of automatic mechanisms, to be actuated from a distance, were constructed by me and exhibited to visitors in my laboratory. In 1896, however, I designed a complete machine capable of a multitude of operations, but the consummation of my labours was delayed until late in 1897. This machine was illustrated and described in my article in the Century Magazine of June 1900, and other periodicals of that time and, when first shown in the beginning of 1898, it created a sensation such as no other invention of mine has ever produced. In November 1898, a basic patent on the novel art was granted to me, but only after the Examiner-in-Chief had come to New York and witnessed the performance, for what I claimed seemed unbelievable. I remember that when later I called on an official in Washington, with a view of offering the invention to the Government, he burst out in laughter upon my telling him what I had accomplished. Nobody thought then that there was the faintest prospect of perfecting such a device. It is unfortunate that in this patent, following the advice of my attorneys, I indicated the control as being affected thru the medium of a single circuit and a well-known form of detector, for the reason that I had not yet secured protection on my methods and apparatus for individualization. As a matter of fact, my boats were controlled thru the joint action of several circuits and interference of every kind was excluded. Most generally I employed receiving circuits in the form of loops, including condensers, because the discharges of my high-tension transmitter ionized the air in the hall so that even a very small aerial would draw electricity from the surrounding atmosphere for hours. Just to give an idea, I found, for instance, that a bulb 12" in diameter, highly exhausted, and with one single terminal to which a short wire was attached, would deliver well on to one thousand successive flashes before all charge of the air in the laboratory was neutralized. The loop form of receiver was not sensitive to such a disturbance and it is curious to note that it is becoming popular at this late date. In reality it collects much less energy than the aerials or a long-grounded wire, but it so happens that it does away with a number of defects inherent to the present wireless devices. In demonstrating my invention before audiences, the visitors were requested to ask any questions, however involved, and the automaton would answer them by signs. This was considered magic at that time but was extremely simple, for it was myself who gave the replies by means of the device.

At the same period another larger telautomatic boat was constructed a photograph of which is shown in this number of the ELECTRICAL EXPERIMENTER. It was controlled by loops, having several turns placed in the hull, which was made entirely water-tight and capable of submergence. The apparatus was similar to that used in the first with the exception of certain special features I introduced as, for example, incandescent lamps which afforded a visible evidence of the proper functioning of the machine.

These automata, controlled within the range of vision of the operator, were, however, the first and rather crude steps in the evolution of the Art of Telautomatics as I had conceived it. The next logical improvement was its application to automatic mechanisms beyond the limits of vision and at great distance from the center of control, and I have ever since advocated their employment as instruments of warfare in preference to guns. The importance of this now seems to be recognized, if I am to judge from casual announcements thru the press of achievements which are said to be extraordinary but contain no merit of novelty, whatever. In an imperfect manner it is practicable, with the existing wireless plants, to launch an aeroplane, have it follow a certain approximate course, and perform some operation at a distance of many hundreds of miles. A machine of this kind can also be mechanically controlled in several ways and I have no doubt that it may prove of some usefulness in war. But there are, to my best knowledge, no instrumentalities in existence today with which such an object could be accomplished in a precise manner. I have devoted years of study to this matter and have evolved means, making such and greater wonders easily realizable.

As stated on a previous occasion, when I was a student at college, I conceived a flying machine quite unlike the present ones. The underlying principle was sound but could not be carried into practice for want of a prime mover of sufficiently great activity. In recent years I have successfully solved this problem and am now planning aerial machines devoid of sustaining planes, ailerons, propellers and other external attachments, which will be capable of immense speeds and are very likely to furnish powerful arguments for peace in the near future. Such a machine, sustained and propelled entirely by reaction, is shown on page 108 and is supposed to be controlled either mechanically or by wireless energy. By installing proper plants, it will be practicable to project a missile of this kind into the air and drop it almost on the very spot designated, which may be thousands of miles away. But we are not going to stop at this. Telautomata will be ultimately produced, capable of acting as if possessed of their own intelligence, and their advent will create a revolution. As early as 1898 I proposed to representatives of a large manufacturing concern the construction and public exhibition of an automobile carriage which, left to itself, would perform a great variety of operations involving something akin to judgment. But my proposal was deemed chimerical at that time, and nothing came from it.

At present many of the ablest minds are trying to devise expedients for preventing a repetition of the awful conflict which is only theoretically ended and the duration and main issues of which I have correctly predicted in an article printed in the Sun of December 20, 1914. The proposed League is not a remedy but on the contrary, in the opinion of a number of competent men, may bring about results just the opposite. It is particularly regrettable that a punitive policy was adopted in framing the terms of peace, because a few years hence it will be possible for nations to fight without armies, ships or guns, by weapons far more terrible, to the destructive action and range of which there is virtually no limit. A city, at any distance whatsoever from the enemy, can be destroyed by him and no power on earth can stop him from doing so. If we want to avert an impending calamity and a state of things which may transform this globe into an inferno, we should push the development of flying machines and wireless transmission of energy without an instant's delay and with all the power and resources of the nation.

# EXPERIMENTS WITH ALTERNATE CURRENTS OF HIGH POTENTIAL AND HIGH FREQUENCY

# Experiments With Alternate Currents Of High Potential And High Frequency

I cannot find words to express how deeply I feel the honor of addressing some of the foremost thinkers of the present time, and so many able scientific men, engineers, and electricians, of the country greatest in scientific achievements.

The results which I have the honor to present before such a gathering I cannot call my own. There are among you not a few who can lay better claim than myself on any feature of merit which this work may contain. I need not mention many names which are world-known names of those among you who are recognized as the leaders in this enchanting science; but one, at least, I must mention a name which could not be omitted in a demonstration of this kind. It is a name associated with the most beautiful invention ever made it is Crookes!

When I was at college, a good time ago, I read, in a translation (for then I was not familiar with your magnificent language), the description of his experiments on radiant matter. I read it only once in my life that time yet every detail about that charming work I can remember this day. Few are the books, let me say, which can make such an impression upon the mind of a student.

But if, on the present occasion, I mention this name as one of many your institution can boast of, it is because I have more than one reason to do so. For what I have to tell you and to show you this evening concerns, in a large measure, that same vague world which Professor Crookes has so ably explored; and, more than this, when I trace back the mental process which led me to these advances which even by myself cannot be considered trifling, since they are so appreciated by you, I believe that their real origin, that which started me to work in this direction and brought me to them, after a long period of constant thought, was that fascinating little book which I read many years ago.

And now that I have made a feeble effort to express my homage and acknowledge my indebtedness to him and others among you, I will make a second effort, which I hope you will not find so feeble as the first, to entertain you.

Give me leave to introduce the subject in a few words.

A short time ago I had the honor to bring before our American Institute of Electrical Engineers [A] some results then arrived at by me in a novel line of work. I need not assure you that the many evidence which I have received that English scientific men and engineers were interested in this work have been for me a great reward and encouragement. I will not dwell upon the experiments already described, except with the view of completing, or more clearly expressing, some ideas advanced by me before, and also with the view of rendering the study here presented self-contained, and my remarks on the subject of this evening's lecture consistent.

This investigation, then, it goes without saying, deals with alternating currents, and to be more precise, with alternating currents of high potential and high frequency. Just in how much a very high frequency is essential for the production of the results presented is a question which even with my present experience, would embarrass me to answer.

---

[A]
For Mr. Tesla's American lecture on this subject see THE ELECTRICAL WORLD of July 11, 1891, and for a report of his French lecture see THE ELECTRICAL WORLD of March 26, 1892.

Some of the experiments may be performed with low frequencies; but very high frequencies are desirable, not only on account of the many effects secured by their use, but also as a convenient means of obtaining, in the induction apparatus employed, the high potentials, which in their turn are necessary to the demonstration of most of the experiments here contemplated.

Of the various branches of electrical investigation, perhaps the most interesting and immediately the most promising is that dealing with alternating currents. The progress in this branch of applied science has been so great in recent years that it justifies the most sanguine hopes. Hardly have we become familiar with one fact, when novel experiences are met with, and new avenues of research are opened. Even at this hour possibilities not dreamed of before are, by the use of these currents, partly realized. As in nature all is ebb and tide, all is wave motion, so it seems that; in all branches of industry alternating currents electric wave motion will have the sway.

One reason, perhaps, why this branch of science is being so rapidly developed is to be found in the interest which is attached to its experimental study. We wind a simple ring of iron with coils; we establish the connections to the generator, and with wonder and delight we note the effects of strange forces which we bring into play, which allow us to transform, to transmit and direct energy at will. We arrange the circuits properly, and we see the mass of iron and wires behave as though it were endowed with life, spinning a heavy armature, through invisible connections, with great speed and power with the energy possibly conveyed from a great distance. We observe how the energy of an alternating current traversing the wire manifests itself not so much in the wire as in the surrounding space in the most surprising manner, taking the forms of heat, light, mechanical energy, and, most surprising of all, even chemical affinity. All these observations fascinate us and fill us with an intense desire to know more about the nature of these phenomena. Each day we go to our work in the hope of discovering, in the hope that someone, no matter who, may find a solution of one of the pending great problems, and each succeeding day we return to our task with renewed ardor; and even if we *are* unsuccessful, our work has not been in vain, for in these strivings, in these efforts, we have found hours of untold pleasure, and we have directed our energies to the benefit of mankind.

We may take at random, if you choose any of the many experiments which may be performed with alternating currents; a few of which only, and by no means the most striking, form the subject of this evening's demonstration: they are all equally interesting, equally inciting to thought.

Here is a simple glass tube from which the air has been partially exhausted. I take hold of it; I bring my body in contact with a wire conveying alternating currents of high potential, and the tube in my hand is brilliantly lighted. In whatever position I may put it, wherever I may move it in space, as far as I can reach, its soft, pleasing light persists with undiminished brightness.

Here is an exhausted bulb suspended from a single wire. Standing on an insulated support. I grasp it, and a platinum button mounted in it is brought to vivid incandescence.

Here, attached to a leading wire, is another bulb, which, as I touch its metallic socket, is filled with magnificent colors of phosphorescent light.

Here still another, which by my fingers' touch casts a shadow the Crookes shadow, of the stem inside of it.

Here, again, insulated as I stand on this platform, I bring my body in contact with one of the terminals of the secondary of this induction coil with the end of a wire many miles long and you see streams of light break forth from its distant end, which is set in violent vibration.

Here, once more, I attach these two plates of wire gauze to the terminals of the coil. I set them a distance apart, and I set the coil to work. You may see a small spark pass between the plates. I insert a thick plate of one of the best dielectrics between them, and instead of rendering altogether impossible, as we are used to expect, I aid the passage of the discharge, which, as I insert the plate, merely changes in appearance, and assumes the form of luminous streams.

Is there, I ask, can there be, a more interesting study than that of alternating currents?

In all these investigations, in all these experiments, which are so very, very interesting, for many years past ever since the greatest experimenter who lectured in this hall discovered its principle, we have had a steady companion, an appliance familiar to everyone, a plaything once, a thing of momentous importance now the induction coil. There is no dearer appliance to the electrician. From the ablest among you, I dare say, down to the inexperienced student, to your lecturer, we all have passed many delightful hours in experimenting with the induction coil. We have watched its play and thought and pondered over the beautiful phenomena which it disclosed to our ravished eyes. So well-known is this apparatus, so familiar are these phenomena to everyone, that my courage nearly fails me when I think that I have ventured to address so able an audience, that I have ventured to entertain you with that same old subject. Here in reality is the same apparatus, and here are the same phenomena, only the apparatus is operated somewhat differently, the phenomena are presented in a different aspect. Some of the results we find as expected, others surprise us, but all captivate our attention, for in scientific investigation each novel result achieved may be the center of a new departure, each novel fact learned may lead to important developments.

Usually in operating an induction coil we have set up a vibration of moderate frequency in the primary, either by means of an interrupter or break, or by the use of an alternator. Earlier English investigators, to mention only Spottiswoode and J.E.H. Gordon, have used a rapid break in connection with the coil. Our knowledge and experience of to-day enables us to see clearly why these coils under the conditions of the tests did not disclose any remarkable phenomena, and why able experimenters failed to perceive many of the curious effects which have since been observed.

In the experiments such as performed this evening, we operate the coil either from a specially constructed alternator capable of giving many thousands of reversals of current per second, or, by disruptively discharging a condenser through the primary, we set up a vibration in the secondary circuit of a frequency of many hundred thousand or millions per second, if we so desire; and in using either of these means we enter a field as yet unexplored.

It is impossible to pursue an investigation in any novel line without finally making some interesting observation or learning some useful fact. That this statement is applicable to the subject of this lecture the many curious and unexpected phenomena which we observe afford a convincing proof. By way of illustration, take for instance the most obvious phenomena, those of the discharge of the induction coil.

Here is a coil which is operated by currents vibrating with extreme rapidity, obtained by disruptively discharging a Leyden jar. It would not surprise a student were the lecturer to say that the secondary of this coil consists of a small length of comparatively stout wire; it would not surprise him were the lecturer to state that, in spite of this, the coil is capable of giving any potential which the best insulation of the turns is able to withstand: but although he may be prepared, and even be indifferent as to the anticipated result, yet the aspect of the discharge of the coil will surprise and interest him. Everyone is familiar with the discharge of an ordinary coil; it need not be reproduced here. But, by way of contrast, here is a form of discharge of a coil, the primary current of which is vibrating several hundred thousand times per second. The discharge of an ordinary coil appears as a simple line or band of light. The discharge of this coil appears in the form of powerful brushes and luminous streams issuing from all points of the two straight wires attached to the terminals of the secondary. (Fig. 1.)

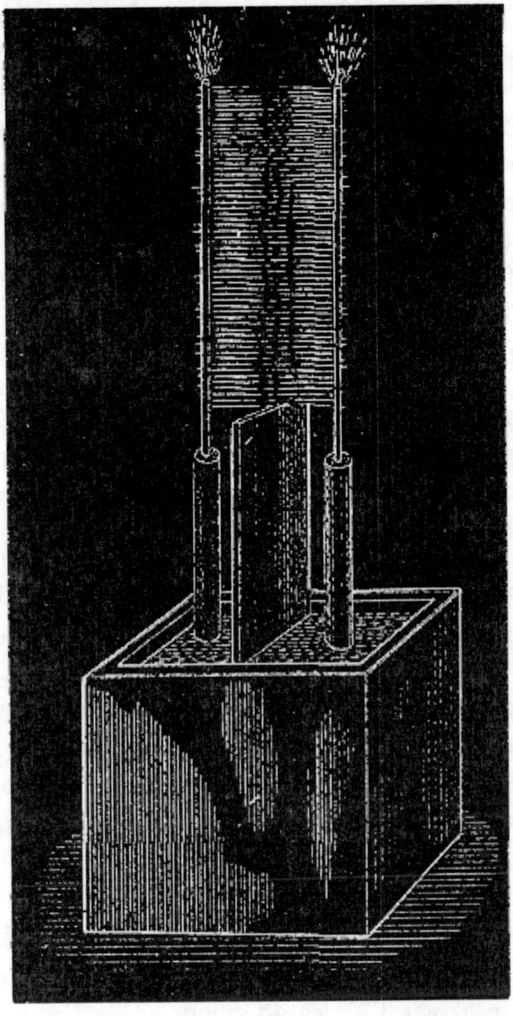

Fig. 1.—Discharge Between Two Wires with Frequencies of a Few Hundred Thousand per Second.

Now compare this phenomenon which you have just witnessed with the discharge of a Holtz or Wimshurst machine that another interesting appliance so dear to the experimenter. What a difference there is between these phenomena! And yet, had I made the necessary arrangements which could have been made easily, were it not that they would interfere with other experiments, I could have produced with this coil sparks which, had I the coil hidden from your view and only two knobs exposed, even the keenest observer among you would find it difficult, if not impossible, to distinguish from those of an influence or friction machine. This may be done in many ways for instance, by operating the induction coil which charges the condenser from an alternating-current machine of very low frequency, and preferably adjusting the discharge circuit so that there are no oscillations set up in it. We then obtain in the secondary circuit, if the knobs are of the required size and properly set, a more or less rapid

succession of sparks of great intensity and small quantity, which possess the same brilliancy, and are accompanied by the same sharp crackling sound, as those obtained from a friction or influence machine.

Another way is to pass through two primary circuits, having a common secondary, two currents of a slightly different period, which produce in the secondary circuit sparks occurring at comparatively long intervals. But, even with the means at hand this evening, I may succeed in imitating the spark of a Holtz machine. For this purpose, I establish between the terminals of the coil which charges the condenser a long, unsteady arc, which is periodically interrupted by the upward current of air produced by it. To increase the current of air I place on each side of the arc, and close to it, a large plate of mica. The condenser charged from this coil discharges into the primary circuit of a second coil through a small air gap, which is necessary to produce a sudden rush of current through the primary. The scheme of connections in the present experiment is indicated in Fig. 2.

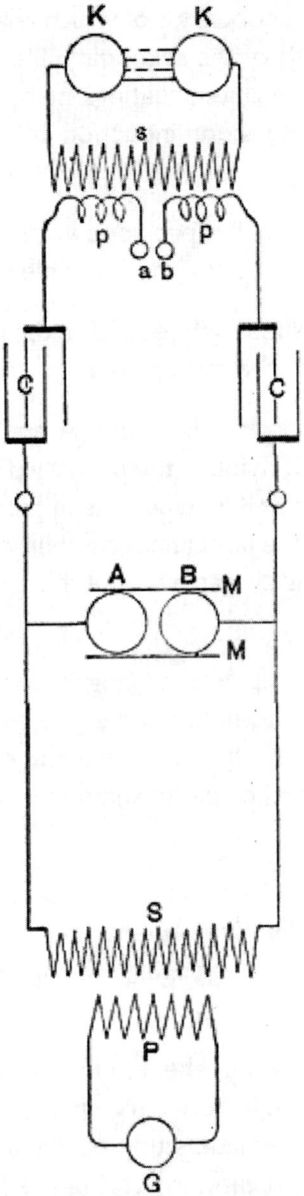


FIG. 2.—IMITATING THE SPARK OF A HOLTZ MACHINE.

G is an ordinarily constructed alternator, supplying the primary P of an induction coil, the secondary S of which charges the condensers or jars C C. The terminals of the secondary are connected to the inside coatings of the jars, the outer coatings being connected to the ends of the primary p p of a second induction coil. This primary p p has a small air gap a b.

The secondary s of this coil is provided with knobs or spheres K K of the proper size and set at a distance suitable for the experiment.

A long arc is established between the terminals A B of the first induction coil. M M are the mica plates.

Each time the arc is broken between A and B the jars are quickly charged and discharged through the primary p p, producing a snapping spark between the knobs K K. Upon the arc forming between A and B the potential falls, and the jars cannot be charged to such high potential as to break through the air gap a b until the arc is again broken by the draught.

In this manner sudden impulses, at long intervals, are produced in the primary p p, which in the secondary s give a corresponding number of impulses of great intensity. If the secondary knobs or spheres, K K, are of the proper size, the sparks show much resemblance to those of a Holtz machine.

But these two effects, which to the eye appear so very different, are only two of the many discharge phenomena. We only need to change the conditions of the test, and again we make other observations of interest.

When, instead of operating the induction coil as in the last two experiments, we operate it from a high frequency alternator, as in the next experiment, a systematic study of the phenomena is rendered much more easy. In such case, in varying the strength and frequency of the currents through the primary, we may observe five distinct forms of discharge, which I have described in my former paper on the subject [A] before the American Institute of Electrical Engineers, May 20, 1891.

It would take too much time, and it would lead us too far from the subject presented this evening, to reproduce all these forms, but it seems to me desirable to show you one of them. It is a brush discharge, which is interesting in more than one respect. Viewed from a near position it resembles much a jet of gas escaping under great pressure. We know that the phenomenon is due to the agitation of the molecules near the terminal, and we anticipate that some heat must be developed by the impact of the molecules against the terminal or against each other. Indeed, we find that the brush is hot, and only a little thought leads us to the conclusion that, could we but reach sufficiently high frequencies, we could produce a brush which would give intense light and heat, and which would resemble in every particular an ordinary flame, save, perhaps, that both phenomena might not be due to the same agent save, perhaps, that chemical affinity might not be *electrical* in its nature.

As the production of heat and light is here due to the impact of the molecules, or atoms of air, or something else besides, and, as we can augment the energy simply by raising the potential, we might, even with frequencies obtained from a dynamo machine, intensify the action to such a degree as to bring the terminal to melting heat. But with such low frequencies we would have to deal always with something of the nature of an electric current. If I approach a conducting object to the brush, a thin little spark passes, yet, even with the frequencies used this evening, the tendency to spark is not very great. So, for instance, if I hold a metallic sphere at some distance above the terminal you may see the whole space between the terminal and sphere illuminated by the streams without the spark passing; and with the much higher frequencies obtainable by the disruptive discharge of a condenser, were it not for the sudden impulses, which are comparatively few

---

[A]
See THE ELECTRICAL WORLD, July 11, 1891.

in number, sparking will not occur even at very small distances. However, with incomparably higher frequencies, which we may yet find means to produce efficiently, and provided that electric impulses of such high frequencies could be transmitted through a conductor, the electrical characteristics of the brush discharge would completely vanish no spark would pass, no shock would be felt yet we would still have to deal with an *electric* phenomenon, but in the broad, modern interpretation of the word. In my first paper before referred to, I have pointed out the curious properties of the brush, and described the best manner of producing it, but I have thought it worthwhile to endeavor to express myself more clearly in regard to this phenomenon, because of its absorbing interest.

When a coil is operated with currents of very high frequency, beautiful brush effects may be produced, even if the coil be of comparatively small dimensions. The experimenter may vary them in many ways, and, if it were nothing else, they afford a pleasing sight. What adds to their interest is that they may be produced with one single terminal as well as with two in fact, often better with one than with two.

But of all the discharge phenomena observed, the most pleasing to the eye, and the most instructive, are those observed with a coil which is operated by means of the disruptive discharge of a condenser. The power of the brushes, the abundance of the sparks, when the conditions are patiently adjusted, is often amazing. With even a very small coil, if it be so well insulated as to stand a difference of potential of several thousand volts per turn, the sparks may be so abundant that the whole coil may appear a complete mass of fire.

Curiously enough the sparks, when the terminals of the coil are set at a considerable distance, seem to dart in every possible direction as though the terminals were perfectly independent of each other. As the sparks would soon destroy the insulation it is necessary to prevent them. This is best done by immersing the coil in a good liquid insulator, such as boiled-out oil. Immersion in a liquid may be considered almost an absolute necessity for the continued and successful working of such a coil.

It is of course out of the question, in an experimental lecture, with only a few minutes at disposal for the performance of each experiment, to show these discharge phenomena to advantage, as to produce each phenomenon at its best a very careful adjustment is required. But even if imperfectly produced, as they are likely to be this evening, they are sufficiently striking to interest an intelligent audience.

Before showing some of these curious effects I must, for the sake of completeness, give a short description of the coil and other apparatus used in the experiments with the disruptive discharge this evening.

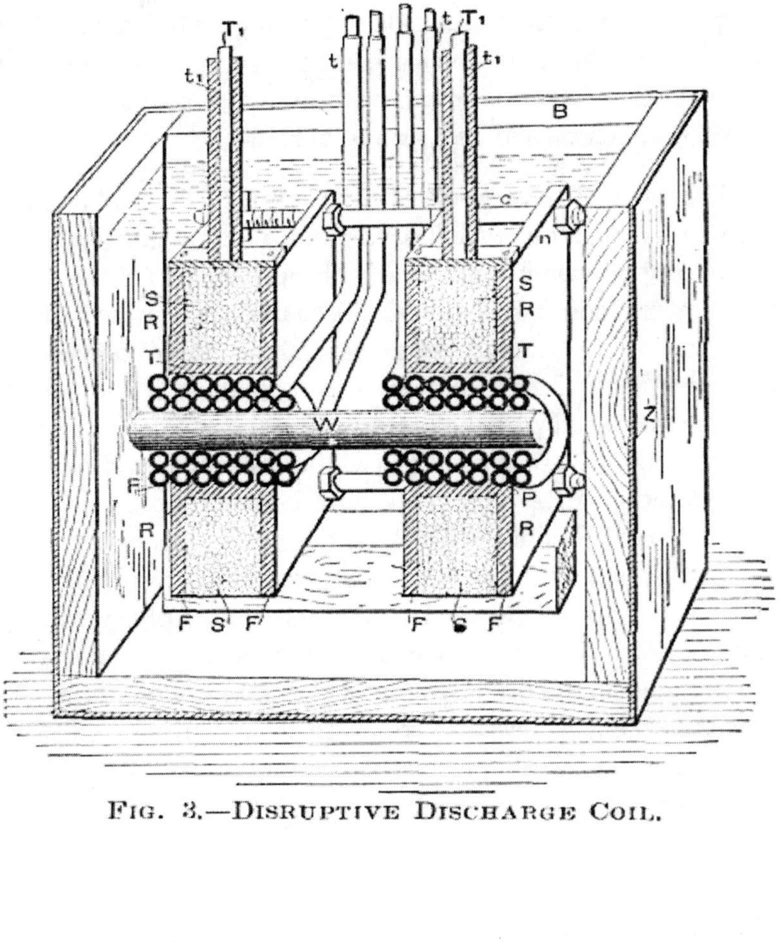


FIG. 3.—DISRUPTIVE DISCHARGE COIL.

It is contained in a box B (Fig. 3) of thick boards of hard wood, covered on the outside with zinc sheet Z, which is carefully soldered all around. It might be advisable, in a strictly scientific investigation, when accuracy is of great importance, to do away with the metal cover, as it might introduce many errors, principally on account of its complex action upon the coil, as a condenser of very small capacity and as an electrostatic and electromagnetic screen. When the coil is used for such experiments as are here contemplated, the employment of the metal cover offers some practical advantages, but these are not of sufficient importance to be dwelt upon.

The coil should be placed symmetrically to the metal cover, and the space between should, of course, not be too small, certainly not less than, say, five centimeters, but much more if possible; especially the two sides of the zinc box, which are at right angles to the axis of the coil, should be sufficiently remote from the latter, as otherwise they might impair its action and be a source of loss.

The coil consists of two spools of hard rubber R R, held apart at a distance of 10 centimeters by bolts c and nuts n, likewise of hard rubber. Each spool comprises a tube T of approximately 8 centimeters inside diameter, and 3 millimeters thick, upon which are screwed two flanges F F, 24 centimeters square, the space between the flanges being about 3 centimeters. The secondary, S S, of the best gutta percha-covered wire, has 26 layers, 10 turns in each, giving for each half a total of 260 turns. The two halves are wound oppositely and connected in series, the connection between both being made over the primary. This disposition, besides being convenient, has the advantage that when the coil is well balanced that is, when *both* of its terminals T1 T1 are connected to bodies or devices of equal capacity there is not much danger of breaking through to the primary, and the insulation between the primary and the secondary need not be thick. In using the coil, it is advisable to attach to both terminals' devices of nearly equal capacity, as, when the capacity of the terminals is not equal, sparks will be apt to pass to the primary. To avoid this, the middle point of the secondary may be connected to the primary, but this is not always practicable.

The primary P P is wound in two parts, and oppositely, upon a wooden spoon W, and the four ends are led out of the oil through hard rubber tubes t t. The ends of the secondary T1 T1 are also led out of the oil through rubber tubes t1 t1 of great thickness. The primary and secondary layers are insulated by cotton cloth, the thickness of the insulation, of course, bearing some proportion to the difference of potential between the turns of the different layers. Each half of the primary has four layers, 24 turns in each, this giving a total of 96 turns. When both the parts are connected in series, this gives a ratio of conversion of about 1:2.7, and with the primaries in multiple, 1:5.4; but in operating with very rapidly alternating currents this ratio does not convey even an approximate idea of the ratio of the E.M.Fs. in the primary and secondary circuits. The coil is held in position in the oil on wooden supports, there being about 5 centimeters thickness of oil all round. Where the oil is not specially needed, the space is filled with pieces of wood, and for this purpose principally the wooden box B surrounding the whole is used.

The construction here shown is, of course, not the best on general principles, but I believe it is a good and convenient one for the production of effects in which an excessive potential and a very small current are needed.

In connection with the coil, I use either the ordinary form of discharger or a modified form. In the former I have introduced two changes which secure some advantages, and which are obvious. If they are mentioned, it is only in the hope that some experimenter may find them of use.

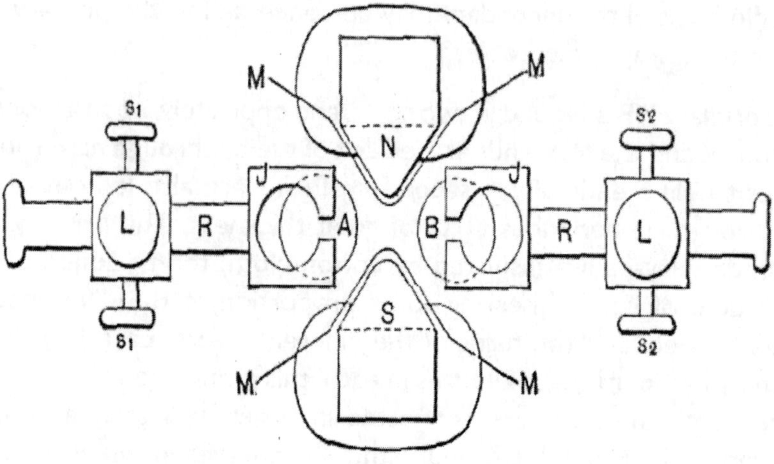


FIG. 4.—ARRANGEMENT OF IMPROVED DISCHARGER AND MAGNET.

One of the changes is that the adjustable knobs A and B (Fig. 4), of the discharger are held in jaws of brass, J J, by spring pressure, this allowing of turning them successively into different positions, and so doing away with the tedious process of frequent polishing up.

The other change consists in the employment of a strong electromagnet N S, which is placed with its axis at right angles to the line joining the knobs A and B and produces a strong magnetic field between them. The pole pieces of the magnet are movable and properly formed so as to protrude between the brass knobs, in order to make the field as intense as possible; but to prevent the discharge from jumping to the magnet the pole pieces are protected by a layer of mica, M M, of sufficient thickness. s1 s1 and s2 s2 are screws for fastening the wires. On each side one of the screws is for large and the other for small wires. L L are screws for fixing in position the rods R R, which support the knobs.

In another arrangement with the magnet, I take the discharge between the rounded pole pieces themselves, which in such case are insulated and preferably provided with polished brass caps.

The employment of an intense magnetic field is of advantage principally when the induction coil or transformer which charges the condenser is operated by currents of very low frequency. In such a case the number of the fundamental discharges between the knobs may be so small as to render the currents produced in the secondary unsuitable for many experiments. The intense magnetic field then serves to blow out the arc between the knobs as soon as it is formed, and the fundamental discharges occur in quicker succession.

Instead of the magnet, a draught or blast of air may be employed with some advantage. In this case the arc is preferably established between the knobs A B, in Fig. 2 (the knobs a b being generally joined, or entirely done away with), as in this disposition the arc is long and unsteady and is easily affected by the draught.

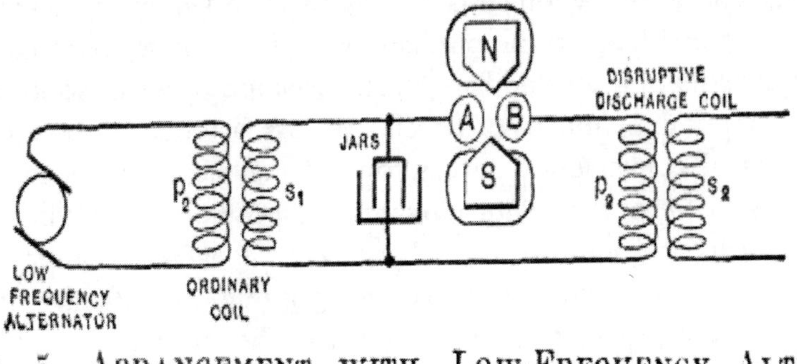


FIG. 5.—ARRANGEMENT WITH LOW-FREQUENCY ALTERNATOR AND IMPROVED DISCHARGER.

When a magnet is employed to break the arc, it is better to choose the connection indicated diagrammatically in Fig. 5, as in this case the currents forming the arc are much more powerful, and the magnetic field exercises a greater influence. The use of the magnet permits, however, of the arc being replaced by a vacuum tube, but I have encountered great difficulties in working with an exhausted tube.

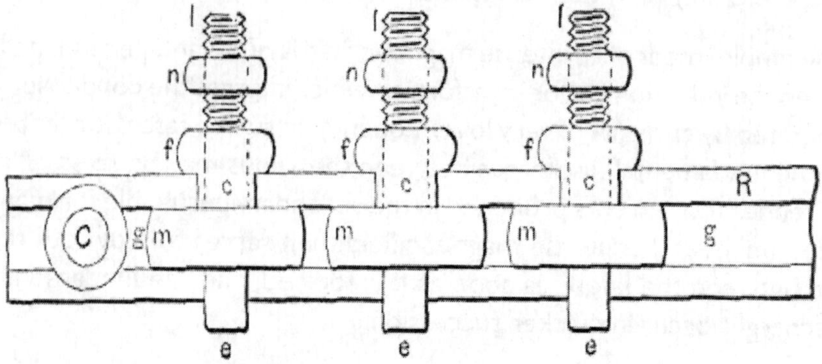


FIG. 6.—DISCHARGER WITH MULTIPLE GAPS.

The other form of discharger used in these and similar experiments is indicated in Figs. 6 and 7. It consists of a number of brass pieces c c (Fig. 6), each of which comprises a spherical middle portion m with an extension e below which is merely used to fasten the piece in a lathe when polishing up the discharging surface and a column above, which consists of a knurled flange f surmounted by a threaded stem l carrying a nut n, by means of which a wire is fastened to the column. The flange f conveniently serves for holding the brass piece when fastening the wire, and for turning it in any position when it becomes necessary to present a fresh discharging surface. Two stout strips of hard rubber R R, with planed grooves g g (Fig. 7) to fit the middle portion of the pieces c c, serve to clamp the latter and hold them firmly in position by means of two bolts C C (of which only one is shown) passing through the ends of the strips.

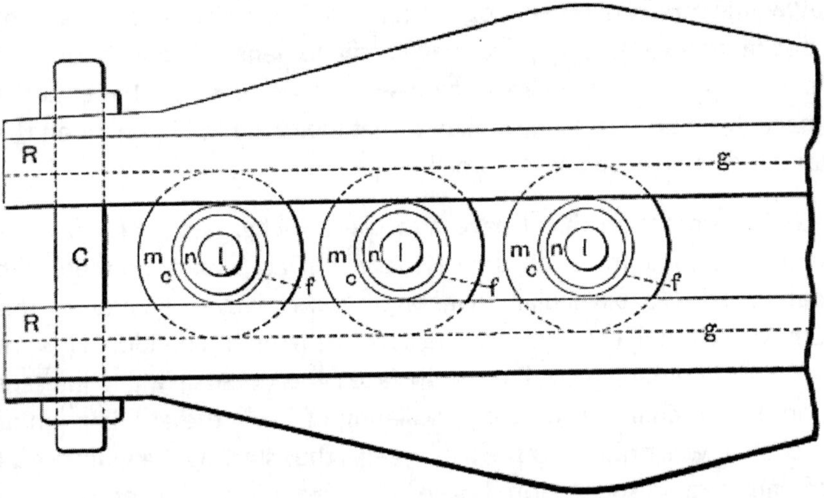


FIG. 7.—DISCHARGER WITH MULTIPLE GAPS.

In the use of this kind of discharger I have found three principal advantages over the ordinary form. First, the dielectric strength of a given total width of air space is greater when a great many small air gaps are used instead of one, which permits of working with a smaller length of air gap, and that means smaller loss and less deterioration of the metal; secondly by reason of splitting the arc up into smaller arcs, the polished surfaces are made to last much longer; and, thirdly, the apparatus affords some gauge in the experiments. I usually set the pieces by putting between them sheets of uniform thickness at a certain very small distance which is known from the experiments of Sir William Thomson to require a certain electromotive force to be bridged by the spark.

It should, of course, be remembered that the sparking distance is much diminished as the frequency is increased. By taking any number of spaces the experimenter has a rough idea of the electromotive force, and he finds it easier to repeat an experiment, as he has not the trouble of setting the knobs again and again. With this kind of discharger, I have been able to maintain an oscillating motion without any spark being visible with the naked eye between the knobs, and they would not show a very appreciable rise in temperature. This form of discharge also lends itself to many arrangements of condensers and circuits which are often very convenient and timesaving. I have used it preferably in a disposition similar to that indicated in Fig. 2, when the currents forming the arc are small.

I may here mention that I have also used dischargers with single or multiple air gaps, in which the discharge surfaces were rotated with great speed. No particular advantage was, however, gained by this method, except in cases where the currents from the condenser were large and the keeping cool of the surfaces was necessary, and in cases when, the discharge not being oscillating of itself, the arc as soon as established was broken by the air current, thus starting the vibration at intervals in rapid succession. I have also used mechanical interrupters in many ways. To avoid the difficulties with frictional contacts, the preferred plan adopted was to establish the arc and rotate through it at great speed a rim of mica provided with many holes and fastened to a steel plate. It is understood, of course, that the employment of a magnet, air current, or other interrupter, produces no effect worth noticing, unless the self-induction, capacity and resistance are so related that there are oscillations set up upon each interruption.

I will now endeavor to show you some of the most note-worthy of these discharge phenomena.

I have stretched across the room two ordinary cotton covered wires, each about 7 meters in length. They are supported on insulating cords at a distance of about 30 centimeters. I attach now to each of the terminals of the coil one of the wires and set the coil in action. Upon turning the lights off in the room you see the wires strongly illuminated by the streams issuing abundantly from their whole surface in spite of the cotton covering, which may even be very thick. When the experiment is performed under good conditions, the light from the wires is sufficiently intense to allow distinguishing the objects in a room. To produce the best result it is, of course, necessary to carefully adjust the capacity of the jars, the arc between the knobs and the length of the wires. My experience is that calculation of the length of the wires leads, in such case, to no result whatever. The experimenter will do best to take the wires at the start very long, and then adjust by cutting off first long pieces, and then smaller and smaller ones as he approaches the right length.

A convenient way is to use an oil condenser of very small capacity, consisting of two small adjustable metal plates, in connection with this and similar experiments. In such case I take wires rather short and set at the beginning the condenser plates at maximum distance. If the streams for the wires increase by approach of the plates, the length of the wires is about right; if they diminish the wires are too long for that frequency and potential. When a condenser is used in connection with experiments with such a coil, it should be an oil condenser by all means, as in using an air condenser considerable energy might be wasted. The wires leading to the plates in the oil should be very thin, heavily coated with some insulating compound, and provided with a conducting covering, this preferably extending under the surface of the oil. The conducting cover should not be too near the terminals, or ends, of the wire, as a spark would be apt to jump from the wire to it. The conducting coating is used to diminish the air losses, in virtue of its action as an electrostatic screen. As to the size of the vessel containing the oil, and the size of the plates, the experimenter gains at once an idea from a rough trial. The size of the plates in *oil is*, however, calculable, as the dielectric losses are very small.

In the preceding experiment it is of considerable interest to know what relation the quantity of the light emitted bears to the frequency and potential of the electric impulses. My opinion is that the heat as well as light effects produced should be proportionate, under otherwise equal conditions of test, to the product of frequency and square of potential, but the experimental verification of the law, whatever it may be, would be exceedingly difficult. One thing is certain, at any rate, and that is, that in augmenting the potential and frequency we rapidly intensify the streams; and, though it may be very sanguine, it is surely not altogether hopeless to expect that we may succeed in producing a practical illuminant on these lines. We would then be simply using burners or flames, in which there would be no chemical process, no consumption of material, but merely a transfer of energy, and which would, in all probability emit lighter and less heat than ordinary flames.

The luminous intensity of the streams is, of course, considerably increased when they are focused upon a small surface. This may be shown by the following experiment:

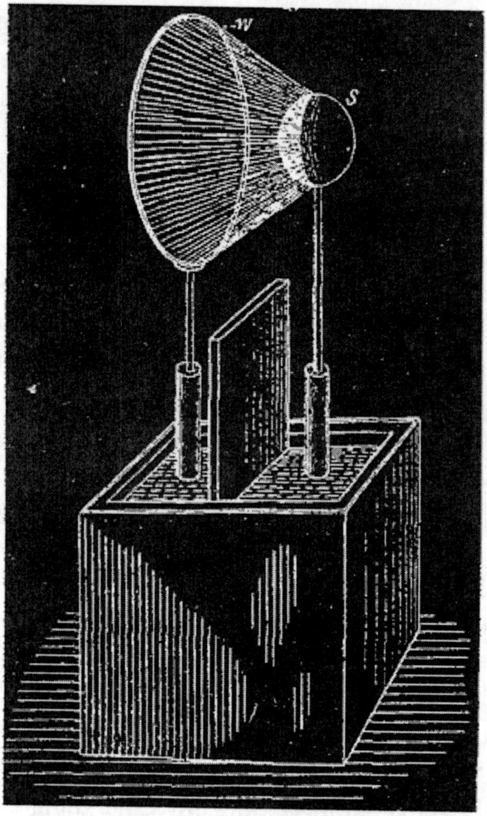


FIG. 8.—EFFECT PRODUCED BY CONCENTRATING STREAMS.

I attach to one of the terminals of the coil a wire w (Fig. 8), bent in a circle of about 30 centimeters in diameter, and to the other terminal I fasten a small brass sphere s, the surface of the wire being preferably equal to the surface of the sphere, and the center of the latter being in a line at right angles to the plane of the wire circle and passing through its center. When the discharge is established under proper conditions, a luminous hollow cone is formed, and in the dark one-half of the brass sphere is strongly illuminated, as shown in the cut.

By some artifice or other, it is easy to concentrate the streams upon small surfaces and to produce very strong light effects. Two thin wires may thus be rendered intensely luminous.

In order to intensify the streams, the wires should be very thin and short; but as in this case their capacity would be generally too small for the coil at least, for such a one as the present, it is necessary to augment the capacity to the required value, while, at the same time, the surface of the wires remains very small. This may be done in many ways.

Here, for instance, I have two plates, R R, of hard rubber (Fig. 9), upon which I have glued two very thin wires w w, so as to form a name. The wires may be bare or covered with the best insulation it is immaterial for the success of the experiment. Well insulated wires, if anything, are preferable. On the back of each plate, indicated by the shaded portion, is a tinfoil coating t t. The plates are placed in line at a sufficient distance to prevent a spark passing from one to the other wire. The two tinfoil coatings I have joined by a conductor C, and the two wires I presently connect to the terminals of the coil. It is now easy, by varying the strength and frequency of the currents through the primary, to find a point at which, the capacity of the system is best suited to the conditions, and the wires become so strongly luminous that, when the light in the room is turned off the name formed by them appears in brilliant letters.

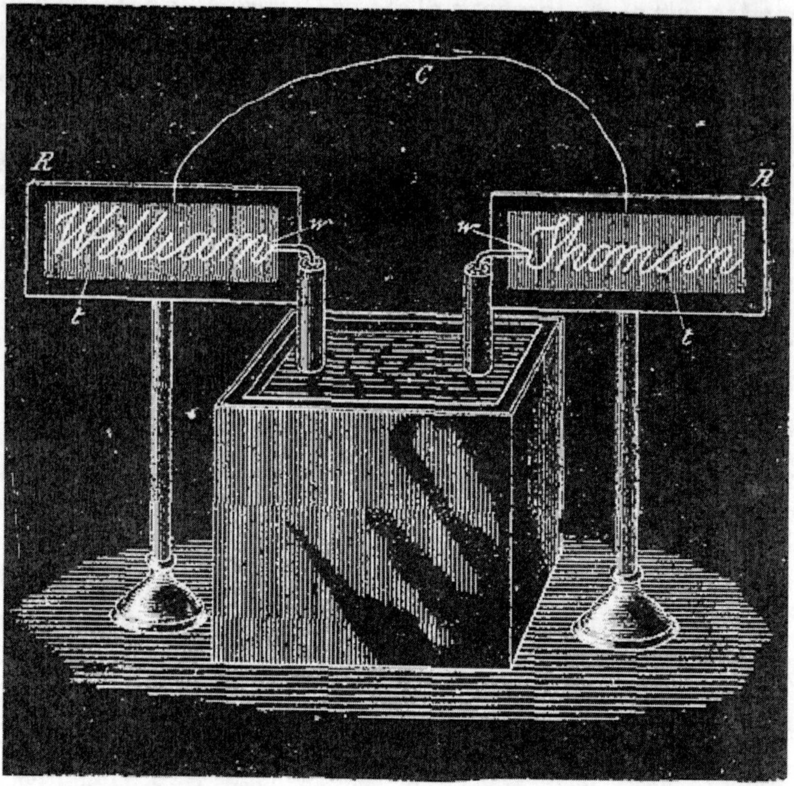


FIG. 9.—WIRES RENDERED INTENSELY LUMINOUS.

It is perhaps preferable to perform this experiment with a coil operated from an alternator of high frequency, as then, owing to the harmonic rise and fall, the streams are very uniform, though they are less abundant then when produced with such a coil as the present. This experiment, however, may be performed with low frequencies, but much less satisfactorily.

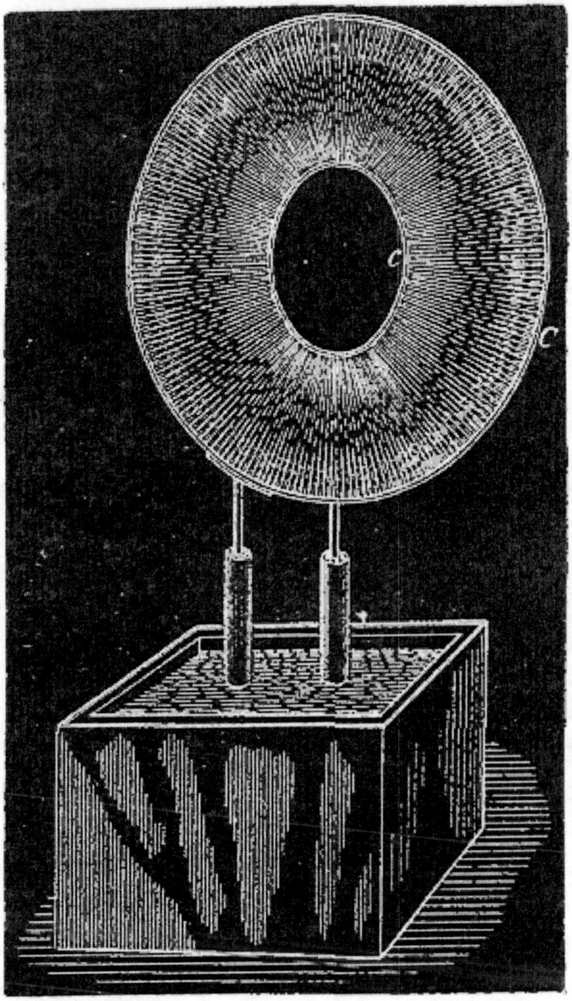


FIG. 10.—LUMINOUS DISCS.

When two wires, attached to the terminals of the coil, are set at the proper distance, the streams between them may be so intense as to produce a continuous luminous sheet. To show this phenomenon I have here two circles, C and c (Fig. 10), of rather stout wire, one being about 80 centimeters and the other 30 centimeters in diameter. To each of the terminals of the coil I attach one of the circles. The supporting wires are so bent that the circles may be placed in the same plane, coinciding as nearly as possible. When the light in the room is turned off and the coil set to work, you see the whole space between the wires uniformly filled with streams, forming a luminous disc, which could be seen from a considerable distance, such is the intensity of the streams. The outer circle could have been much larger than the present one; in fact, with this coil I have used much larger circles, and I have been able to produce a strongly luminous sheet, covering an area of more than one square meter, which is a remarkable effect with this very small coil. To avoid uncertainty, the circle has been taken smaller, and the area is now about 0.43 square meter.

The frequency of the vibration, and the quickness of succession of the sparks between the knobs, affect to a marked degree the appearance of the streams. When the frequency is very low, the air gives way in more or less the same manner, as by a steady difference of potential, and the streams consist of distinct threads, generally mingled with thin sparks, which probably correspond to the successive discharges occurring between the knobs. But when the frequency is extremely high, and the arc of the discharge produces a very *loud* but *smooth* sound showing both that oscillation takes place and that the sparks succeed each other with great rapidity, then the luminous streams formed are perfectly uniform. To reach this result very small coils and jars of small capacity should be used. I take two tubes of thick Bohemian glass, about 5 centimeters in diameter and 20 centimeters long. In each of the tubes I slip a primary of very thick copper wire. On the top of each tube, I wind a secondary of much thinner gutta-percha covered wire. The two secondaries I connect in series, the primaries preferably in multiple arcs. The tubes are then placed in a large glass vessel, at a distance of 10 to 15 centimeters from each other, on insulating supports, and the vessel is filled with boiled out oil, the oil reaching about an inch above the tubes. The free ends of the secondary are lifted out of the oil and placed parallel to each other at a distance of about 10 centimeters. The ends which are scraped should be dipped in the oil. Two four-pint jars joined in series may be used to discharge through the primary. When the necessary adjustments in the length and distance of the wires above the oil and in the arc of discharge are made, a luminous sheet is produced between the wires which is perfectly smooth and texture less, like the ordinary discharge through a moderately exhausted tube.

I have purposely dwelt upon this apparently insignificant experiment. In trials of this kind the experimenter arrives at the startling conclusion that, to pass ordinary luminous discharges through gases, no particular degree of exhaustion is needed, but that the gas may be at ordinary or even greater pressure. To accomplish this, a very high frequency is essential; a high potential is likewise required, but this is a merely incidental necessity. These experiments teach us that, in endeavoring to discover novel methods of producing light by the agitation of atoms, or molecules, of a gas, we need not limit our research to the vacuum tube but may look forward quite seriously to the possibility of obtaining the light effects without the use of any vessel whatever, with air at ordinary pressure.

Such discharges of very high frequency, which render luminous the air at ordinary pressures, we have probably often occasion to witness in Nature. I have no doubt that if, as many believe, the aurora borealis is produced by sudden cosmic disturbances, such as eruptions at the sun's surface, which set the electrostatic charge of the earth in an extremely rapid vibration, the red glow observed is not confined to the upper rarefied strata of the air, but the discharge traverses, by reason of its very high frequency, also the dense atmosphere in the form of a glow, such as we ordinarily produce in a slightly exhausted tube. If the frequency were very low, or even more so, if the charge were not at all vibrating, the dense air would break down as in a lightning discharge. Indications of such breaking down of the lower dense strata of the air have been repeatedly observed at the occurrence of this marvelous phenomenon; but if it does occur, it can only be attributed to the fundamental disturbances, which are few in number, for the vibration produced by them would be far too rapid to allow a disruptive break. It is the original and irregular impulses which affect the instruments; the superimposed vibrations probably pass unnoticed.

When an ordinary low frequency discharge is passed through moderately rarefied air, the air assumes a purplish hue. If by some means or other we increase the intensity of the molecular, or atomic, vibration, the gas changes to a white color. A similar change occurs at ordinary pressures with electric impulses of very high frequency. If the molecules of the air around a wire are moderately agitated, the brush formed is reddish or violet; if the vibration is rendered sufficiently intense, the streams become white. We may accomplish this in various ways. In the experiment before shown with the two wires across the room, I have endeavored to secure the result by pushing to a high value both the frequency and potential: in the experiment with the thin wires glued on the rubber plate I have concentrated the action upon a very small surface in other words, I have worked with a great electric density.

A most curious form of discharge is observed with such a coil when the frequency and potential are pushed to the extreme limit. To perform the experiment, every part of the coil should be heavily insulated, and only two small spheres or, better still, two sharp-edged metal discs (d d, Fig. 11) of no more than a few centimeters in diameter should be exposed to the air. The coil here used is immersed in oil, and the ends of the secondary reaching out of the oil are covered with an air-tight cover of hard rubber of great thickness. All cracks, if there are any, should be carefully stopped up, so that the brush discharge cannot form anywhere except on the small spheres or plates which are exposed to the air. In this case, since there are no large plates or other bodies of capacity attached to the terminals, the coil is capable of an extremely rapid vibration. The potential may be raised by increasing, as far as the experimenter judges proper, the rate of change of the primary current. With a coil not widely differing from the present, it is best to connect the two primaries in multiple arc; but if the secondary should have a much greater number of turns the primaries should preferably be used in series, as otherwise the vibration might be too fast for the secondary. It occurs under these conditions that misty white streams break forth from the edges of the discs and spread-out phantom-like into space.

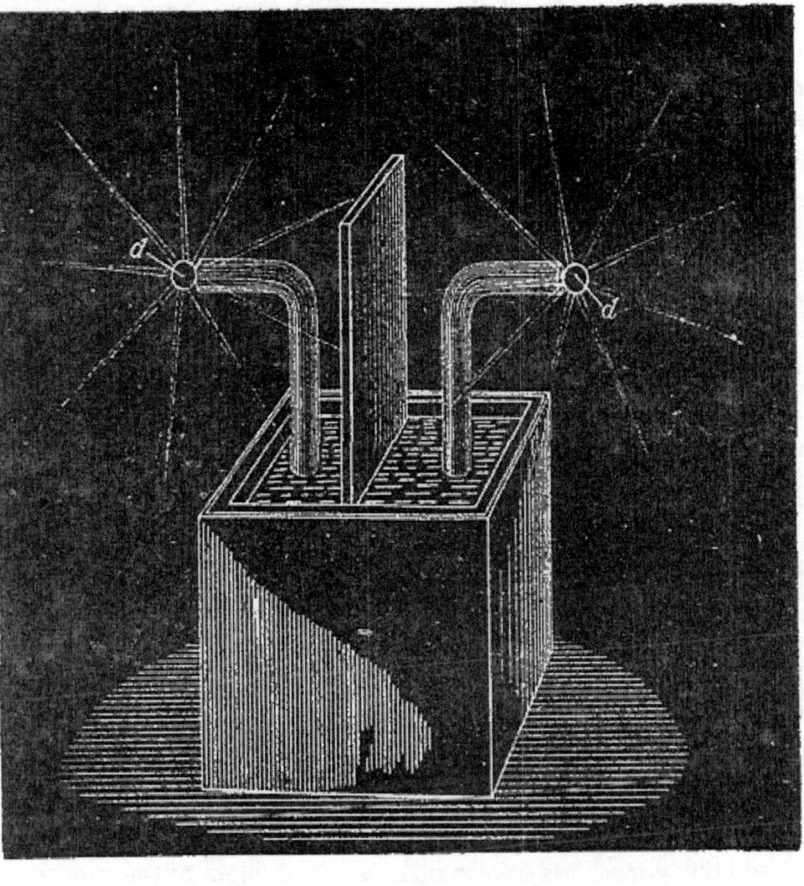


FIG. 11.—PHANTOM STREAMS.

With this coil, when fairly well produced, they are about 25 to 30 centimeters long. When the hand is held against them no sensation is produced, and a spark, causing a shock, jumps from the terminal only upon the hand being brought much nearer. If the oscillation of the primary current is rendered intermittent by some means or other, there is a corresponding throbbing of the streams, and now the hand or other conducting object may be brought in still greater proximity to the terminal without a spark being caused to jump.

Among the many beautiful phenomena which may be produced with such a coil I have here selected only those which appear to possess some features of novelty and lead us to some conclusions of interest. One will not find it at all difficult to produce in the laboratory, by means of it, many other phenomena which appeal to the eye even more than these here shown but present no particular feature of novelty.

Early experimenters describe the display of sparks produced by an ordinary large induction coil upon an insulating plate separating the terminals. Quite recently Siemens performed some experiments in which fine effects were obtained, which were seen by many with interest. No doubt large coils, even if operated with currents of low frequencies, are capable of producing beautiful effects. But the largest coil ever made could not, by far, equal the magnificent display of streams and sparks obtained from such a disruptive discharge coil when properly adjusted. To give an idea, a coil such as the present one will easily cover a plate of 1 meter in diameter completely with the streams. The best way to perform such experiments is to take a very thin rubber or a glass plate and glue on one side of it a narrow ring of tinfoil of very large diameter, and on the other a circular washer, the center of the latter coinciding with that of the ring, and the surfaces of both being preferably equal, so as to keep the coil well balanced. The washer and ring should be connected to the terminals by heavily insulated thin wires. It is easy in observing the effect of the capacity to produce a sheet of uniform streams, or a fine network of thin silvery threads, or a mass of loud brilliant sparks, which completely cover the plate.

Since I have advanced the idea of the conversion by means of the disruptive discharge, in my paper before the American Institute of Electrical Engineers at the beginning of the past year, the interest excited in it has been considerable. It affords us a means for producing any potentials by the aid of inexpensive coils operated from ordinary systems of distribution, and what is perhaps more appreciated, it enables us to convert currents of any frequency into currents of any other lower or higher frequency. But its chief value will perhaps be found in the help which it will afford us in the investigations of the phenomena of phosphorescence, which a disruptive discharge coil is capable of exciting in innumerable cases where ordinary coils, even the largest, would utterly fail.

Considering its probable uses for many practical purposes, and its possible introduction into laboratories for scientific research, a few additional remarks as to the construction of such a coil will perhaps not be found superfluous.

It is, of course, absolutely necessary to employ in such a coil wire provided with the best insulation.

Good coils may be produced by employing wires covered with several layers of cotton, boiling the coil a long time in pure wax, and cooling under moderate pressure. The advantage of such a coil is that it can be easily handled, but it cannot probably give as satisfactory results as a coil immersed in pure oil. Besides, it seems that the presence of a large body of wax affects the coil disadvantageously, whereas this does not seem to be the case with oil. Perhaps it is because the dielectric losses in the liquid are smaller.

I have tried at first silk and cotton covered wires with oil immersion, but I have been gradually led to use gutta-percha covered wires, which proved most satisfactory. Gutta-percha insulation adds, of course, to the capacity of the coil, and this, especially if the coil be large, is a great disadvantage when extreme frequencies are desired; but on the other hand, gutta-percha will withstand much more than an equal thickness of oil, and this advantage should be secured at any price. Once the coil has been immersed, it should never be taken out of the oil for more than a few hours, else the gutta-percha will crack up and the coil will not be worth half as much as before. Gutta-percha is probably slowly attacked by the oil, but after an immersion of eight to nine months I have found no ill effects.

I have obtained in commerce two kinds of gutta-percha wire: in one the insulation sticks tightly to the metal, in the other it does not. Unless a special method is followed to expel all air, it is much safer to use the first kind. I wind the coil within an oil tank so that all interstices are filled up with the oil. Between the layers I use cloth boiled out thoroughly in oil, calculating the thickness according to the difference of potential between the turns. There seems not to be a very great difference whatever kind of oil is used; I use paraffine or linseed oil.

To exclude more perfectly the air, an excellent way to proceed, and easily practicable with small coils, is the following: Construct a box of hard wood of very thick boards which have been for a long time boiled in oil. The boards should be so joined as to safely withstand the external air pressure. The coil being placed and fastened in position within the box, the latter is closed with a strong lid, and covered with closely fitting metal sheets, the joints of which are soldered very carefully. On the top two small holes are drilled, passing through the metal sheet and the wood, and in these holes two small glass tubes are inserted and the joints made airtight. One of the tubes is connected to a vacuum pump, and the other with a vessel containing a sufficient quantity of boiled-out oil. The latter tube has a very small hole at the bottom and is provided with a stopcock. When a fairly good vacuum has been obtained, the stopcock is opened, and the oil slowly fed in. Proceeding in this manner, it is impossible that any big bubbles, which are the principal danger, should remain between the turns. The air is most completely excluded, probably better than by boiling out, which, however, when gutta-percha coated wires are used, is not practicable.

For the primaries I use ordinary line wire with a thick cotton coating. Strands of very thin insulated wires properly interlaced would, of course, be the best to employ for the primaries, but they are not to be had.

In an experimental coil the size of the wires is not of great importance. In the coil here used the primary is No. 12 and the secondary No. 24 Brown & Sharpe gauge wire; but the sections may be varied considerably. It would only imply different adjustments; the results aimed at would not be materially affected.

I have dwelt at some length upon the various forms of brush discharge because, in studying them, we not only observe phenomena which please our eye, but also afford us food for thought, and lead us to conclusions of practical importance. In the use of alternating currents of very high tension, too much precaution cannot be taken to prevent the brush discharge. In a main conveying such currents, in an induction coil or transformer, or in a condenser, the brush discharge is a source of great danger to the insulation. In a condenser especially the gaseous matter must be most carefully expelled, for in it the charged surfaces are near each other, and if the potentials are high, just as sure as a weight will fall if let go, so the insulation will give way if a single gaseous bubble of some size be present, whereas, if all gaseous matter were carefully excluded, the condenser would safely withstand a much higher difference of potential. A main conveying alternating currents of very high tension may be injured merely by a blow hole or small crack in the insulation, the more so as a blowhole is apt to contain gas at low pressure; and as it appears almost impossible to completely obviate such little imperfections, I am led to believe that in our future distribution of electrical energy by currents of very high-tension liquid insulation will be used. The cost is a great drawback, but if we employ an oil as an insulator the distribution of electrical energy with something like 100,000 volts, and even more, become, at least with higher frequencies, so easy that they could be hardly called engineering feats. With oil insulation and alternate current motors transmissions of power can be affected with safety and upon an industrial basis at distances of as much as a thousand miles.

A peculiar property of oils, and liquid insulation in general, when subjected to rapidly changing electric stresses, is to disperse any gaseous bubbles which may be present, and diffuse them through its mass, generally long before any injurious break can occur. This feature may be easily observed with an ordinary induction coil by taking the primary out, plugging up the end of the tube upon which the secondary is wound, and filling it with some fairly transparent insulator, such as paraffine oil. A primary of a diameter something like six millimeters smaller than the inside of the tube may be inserted in the oil. When the coil is set to work one may see, looking from the top through the oil, many luminous points air bubbles which are caught by inserting the primary, and which are rendered luminous in consequence of the violent bombardment. The occluded air, by its impact against the oil, heats it; the oil begins to circulate, carrying some of the air along with it, until the bubbles are dispersed, and the luminous points disappear. In this manner, unless large bubbles are occluded in such way that circulation is rendered impossible, a damaging break is averted, the only effect being a moderate warming up of the oil. If, instead of the liquid, a solid insulation, no matter how thick, were used, a breaking through and injury of the apparatus would be inevitable.

The exclusion of gaseous matter from any apparatus in which the dielectric is subjected to more or less rapidly changing electric forces is, however, not only desirable in order to avoid a possible injury of the apparatus, but also on account of economy. In a condenser, for instance, as long as only a solid or only a liquid dielectric is used, the loss is small; but if a gas under ordinary or small pressure be present the loss may be very great. Whatever the nature of the force acting in the dielectric may be, it seems that in a solid or liquid the molecular displacement produced by the force is small; hence the product of force and displacement is insignificant, unless the force be very great; but in a gas the displacement, and therefore this product, is considerable; the molecules are free to move, they reach high speeds, and the energy of their impact is lost in heat or otherwise. If the gas be strongly compressed, the displacement due to the force is made smaller, and the losses are reduced.

In most of the succeeding experiments I prefer, chiefly on account of the regular and positive action, to employ the alternator before referred to. This is one of the several machines constructed by me for the purposes of these investigations. It has 384 pole projections and is capable of giving currents of a frequency of about 10,000 per second. This machine has been illustrated and briefly described in my first paper before the American Institute of Electrical Engineers, May 20, 1891, to which I have already referred. A more detailed description, sufficient to enable any engineer to build a similar machine, will be found in several electrical journals of that period.

The induction coils operated from the machine are rather small, containing from 5,000 to 15,000 turns in the secondary. They are immersed in boiled-out linseed oil, contained in wooden boxes covered with zinc sheet.

I have found it advantageous to reverse the usual position of the wires, and to wind, in these coils, the primaries on the top; this allowing the use of a much bigger primary, which, of course, reduces the danger of overheating and increases the output of the coil. I make the primary on each side at least one centimeter shorter than the secondary, to prevent the breaking through on the ends, which would surely occur unless the insulation on the top of the secondary be very thick, and this, of course, would be disadvantageous.

When the primary is made movable, which is necessary in some experiments, and many times convenient for the purposes of adjustment, I cover the secondary with wax, and turn it off in a lathe to a diameter slightly smaller than the inside of the primary coil. The latter I provide with a handle reaching out of the oil, which serves to shift it in any position along the secondary.

I will now venture to make, in regard to the general manipulation of induction coils, a few observations bearing upon points which have not been fully appreciated in earlier experiments with such coils and are even now often overlooked.

The secondary of the coil possesses usually such a high self-induction that the current through the wire is inappreciable and may be so even when the terminals are joined by a conductor of small resistance. If capacity is added to the terminals, the self-induction is counteracted, and a stronger current is made to flow through the secondary, though its terminals are insulated from each other. To one entirely unacquainted with the properties of alternating currents nothing will look more puzzling. This feature was illustrated in the experiment performed at the beginning with the top plates of wire gauze attached to the terminals and the rubber plate. When the plates of wire gauze were close together, and a small arc passed between them, the arc *prevented* a strong current from passing through the secondary, because it did away with the capacity on the terminals; when the rubber plate was inserted between, the capacity of the condenser formed counteracted the self-induction of the secondary, a stronger current passed now, the coil performed more work, and the discharge was by far more powerful.

The first thing, then, in operating the induction coil is to combine capacity with the secondary to overcome the self-induction. If the frequencies and potentials are very high gaseous matter should be carefully kept away from the charged surfaces. If Leyden jars are used, they should be immersed in oil, as otherwise considerable dissipation may occur if the jars are greatly strained. When high frequencies are used, it is of equal importance to combine a condenser with the primary. One may use a condenser connected to the ends of the primary or to the terminals of the alternator, but the latter is not to be recommended, as the machine might be injured. The best way is undoubtedly to use the condenser in series with the primary and with the alternator, and to adjust its capacity so as to annul the self-induction of both the latter. The condenser should be adjustable by very small steps, and for a finer adjustment a small oil condenser with movable plates may be used conveniently.

I think it best at this juncture to bring before you a phenomenon, observed by me some time ago, which to the purely scientific investigator may perhaps appear more interesting than any of the results which I have the privilege to present to you this evening.

It may be quite properly ranked among the brush phenomena in fact, it is a brush, formed at, or near, a single terminal in high vacuum.

In bulbs provided with a conducting terminal, though it be of aluminum, the brush has but an ephemeral existence, and cannot, unfortunately, be indefinitely preserved in its most sensitive state, even in a bulb devoid of any conducting electrode. In studying the phenomenon, by all means a bulb having no leading-in wire should be used. I have found it best to use bulbs constructed as indicated in Figs. 12 and 13.

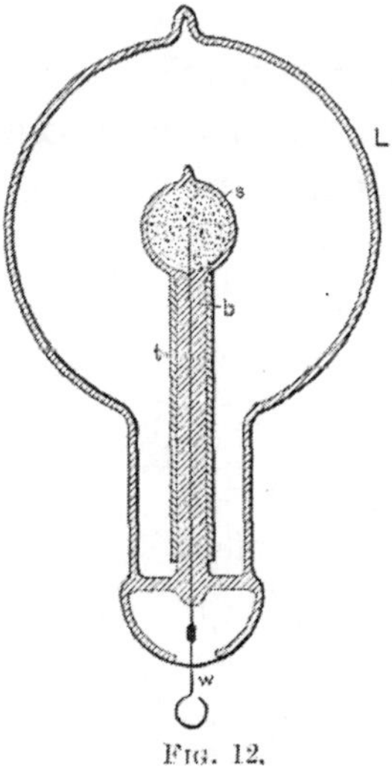


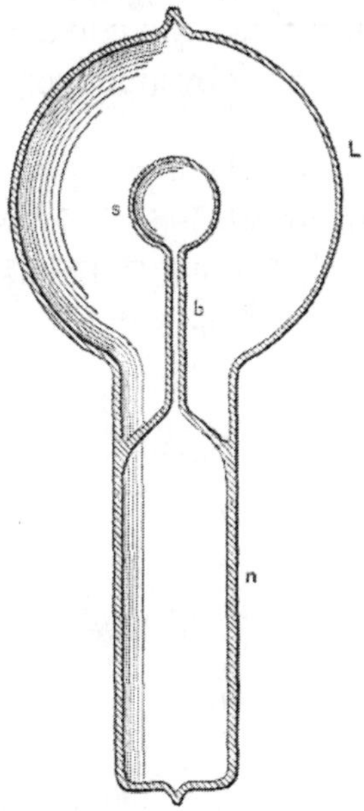


FIG. 12. FIG. 13.
BULBS FOR PRODUCING ROTATING BRUSH.

In Fig. 12 the bulb comprises an incandescent lamp globe L, in the neck of which is sealed a barometer tube b, the end of which is blown out to form a small sphere s. This sphere should be sealed as closely as possible in the center of the large globe. Before sealing, a thin tube t, of aluminum sheet, may be slipped in the barometer tube, but it is not important to employ it.

The small hollow sphere s is filled with some conducting powder, and a wire w is cemented in the neck for the purpose of connecting the conducting powder with the generator.

The construction shown in Fig. 13 was chosen in order to remove from the brush any conducting body which might possibly affect it. The bulb consists in this case of a lamp globe L, which has a neck n, provided with a tube b and small sphere s, sealed to it, so that two entirely independent compartments are formed, as indicated in the drawing. When the bulb is in use, the neck n is provided with a tinfoil coating, which is connected to the generator and acts inductively upon the moderately rarefied and highly conducting gas enclosed in the neck. From there the current passes through the tube b into the small sphere s to act by induction upon the gas contained in the globe L.

It is of advantage to make the tube t very thick, the hole through it very small, and to blow the sphere s very thin. It is of the greatest importance that the sphere s be placed in the center of the globe L.

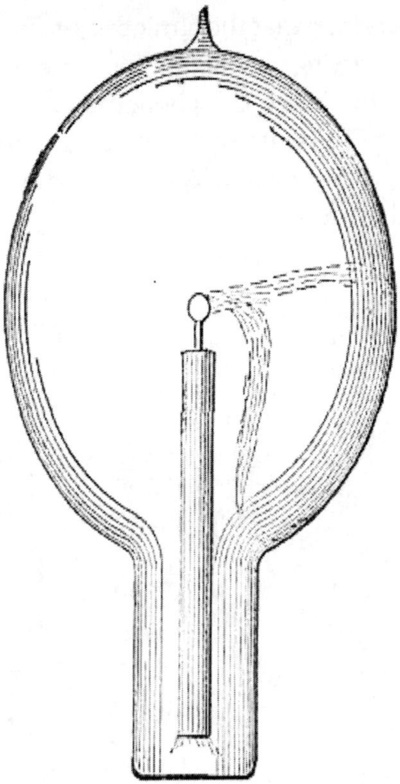

FIG. 14.—FORMS AND PHASES OF THE ROTATING BRUSH.

Figs. 14, 15 and 16 indicate different forms, or stages, of the brush. Fig. 14 shows the brush as it first appears in a bulb provided with a conducting terminal; but, as in such a bulb it very soon disappears often after a few minutes I will confine myself to the description of the phenomenon as seen in a bulb without conducting electrode. It is observed under the following conditions:

When the globe L (Figs. 12 and 13) is exhausted to a very high degree, generally the bulb is not excited upon connecting the wire w (Fig. 12) or the tinfoil coating of the bulb (Fig. 13) to the terminal of the induction coil. To excite it, it is usually sufficient to grasp the globe L with the hand. An intense phosphorescence then spreads at first over the globe, but soon gives place to a white, misty light. Shortly afterward one may notice that the luminosity is unevenly distributed in the globe, and after passing the current for some time the bulb appears as in Fig. 15. From this stage the phenomenon will gradually pass to that indicated in Fig. 16, after some minutes, hours, days or weeks, according as the bulb is worked. Warming the bulb or increasing the potential hastens the transit.

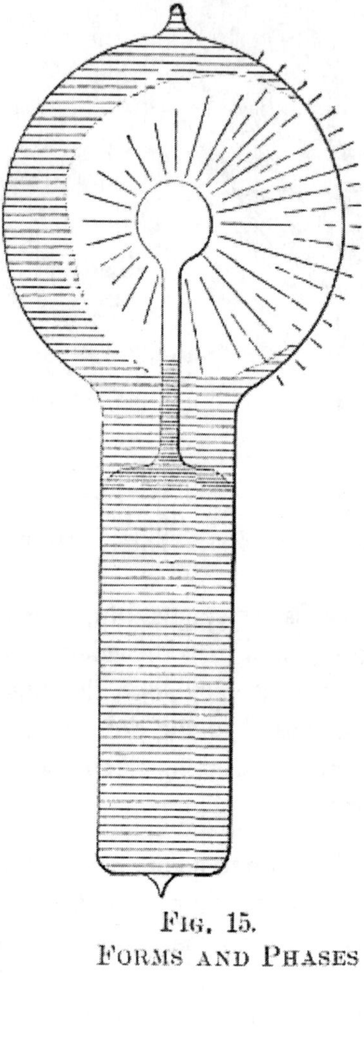

Fig. 15.

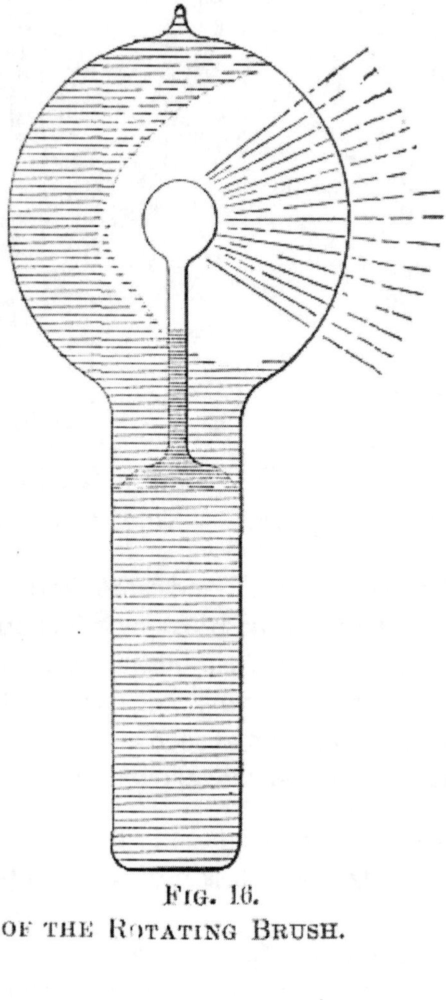

Fig. 16.

Forms and Phases of the Rotating Brush.

When the brush assumes the form indicated in Fig. 16, it may be brought to a state of extreme sensitiveness to electrostatic and magnetic influence. The bulb hanging straight down from a wire, and all objects being remote from it, the approach of the observer at a few paces from the bulb will cause the brush to fly to the opposite side, and if he walks around the bulb, it will always keep on the opposite side. It may begin to spin around the terminal long before it reaches that sensitive stage. When it begins to turn around principally, but also before, it is affected by a magnet, and at a certain stage it is susceptible to magnetic influence to an astonishing degree. A small permanent magnet, with its poles at a distance of no more than two centimeters, will affect it visibly at a distance of two meters, slowing down or accelerating the rotation according to how it is held relatively to the brush. I think I have observed that at the stage when it is most sensitive to magnetic, it is not most sensitive to electrostatic, influence. My explanation is, that the electrostatic attraction between the brush and the glass of the bulb, which retards the rotation, grows much quicker than the magnetic influence when the intensity of the stream is increased.

When the bulb hangs with the globe L down, the rotation is always clockwise. In the southern hemisphere it would occur in the opposite direction and on the equator the brush should not turn at all. The rotation may be reversed by a magnet kept at some distance. The brush rotates best, seemingly, when it is at right angles to the lines of force of the earth. It very likely rotates, when at its maximum speed, in synchronism with the alternations, say 10,000 times a second. The rotation can be slowed down or accelerated by the approach or receding of the observer, or any conducting body, but it cannot be reversed by putting the bulb in any position. When it is in the state of the highest sensitiveness and the potential or frequency be varied the sensitiveness is rapidly diminished. Changing either of these but little will generally stop the rotation. The sensitiveness is likewise affected by the variations of temperature. To attain great sensitiveness, it is necessary to have the small sphere s in the center of the globe L, as otherwise the electrostatic action of the glass of the globe will tend to stop the rotation. The sphere s should be small and of uniform thickness; any dissymmetry of course has the effect to diminish the sensitiveness.

The fact that the brush rotates in a definite direction in a permanent magnetic field seems to show that in alternating currents of very high frequency the positive and negative impulses are not equal, but that one always preponderates over the other.

Of course, this rotation in one direction may be due to the action of two elements of the same current upon each other, or to the action of the field produced by one of the elements upon the other, as in a series motor, without necessarily one impulse being stronger than the other. The fact that the brush turns, as far as I could observe, in any position, would speak for this view. In such case it would turn at any point of the earth's surface. But, on the other hand, it is then hard to explain why a permanent magnet should reverse the rotation, and one must assume the preponderance of impulses of one kind.

As to the causes of the formation of the brush or stream, I think it is due to the electrostatic action of the globe and the dissymmetry of the parts. If the small bulb s and the globe L were perfect concentric spheres, and the glass throughout of the same thickness and quality, I think the brush would not form, as the tendency to pass would be equal on all sides. That the formation of the stream is due to an irregularity is apparent from the fact that it has the tendency to remain in one position, and rotation occurs most generally only when it is brought out of this position by electrostatic or magnetic influence. When in an extremely sensitive state it rests in one position, most curious experiments may be performed with it. For instance, the experimenter may, by selecting a proper position, approach the hand at a certain considerable distance to the bulb, and he may cause the brush to pass off by merely stiffening the muscles of the arm. When it begins to rotate slowly, and the hands are held at a proper distance, it is impossible to make even the slightest motion without producing a visible effect upon the brush. A metal plate connected to the other terminal of the coil affects it at a great distance, slowing down the rotation often to one turn a second.

I am firmly convinced that such a brush, when we learn how to produce it properly, will prove a valuable aid in the investigation of the nature of the forces acting in an electrostatic or magnetic field. If there is any motion which is measurable going on in the space, such a brush ought to reveal it. It is, so to speak, a beam of light, frictionless, devoid of inertia.

I think that it may find practical applications in telegraphy. With such a brush it would be possible to send dispatches across the Atlantic, for instance, with any speed, since its sensitiveness may be so great that the slightest changes will affect it. If it were possible to make the stream more intense and very narrow, its deflections could be easily photographed.

I have been interested to find whether there is a rotation of the stream itself, or whether there is simply a stress traveling around in the bulb. For this purpose, I mounted a light mica fan so that its vanes were in the path of the brush. If the stream itself was rotating the fan would be spun around. I could produce no distinct rotation of the fan, although I tried the experiment repeatedly; but as the fan exerted a noticeable influence on the stream, and the apparent rotation of the latter was, in this case, never quite satisfactory, the experiment did not appear to be conclusive.

I have been unable to produce the phenomenon with the disruptive discharge coil, although every other of these phenomena can be well produced by it many, in fact, much better than with coils operated from an alternator.

It may be possible to produce the brush by impulses of one direction, or even by a steady potential, in which case it would be still more sensitive to magnetic influence.

In operating an induction coil with rapidly alternating currents, we realize with astonishment, for the first time, the great importance of the relation of capacity, self-induction and frequency as regards the general result. The effects of capacity are the most striking, for in these experiments, since the self-induction and frequency both are high, the critical capacity is very small, and need be but slightly varied to produce a very considerable change. The experimenter may bring his body in contact with the terminals of the secondary of the coil or attach to one or both terminals insulated bodies of very small bulk, such as bulbs, and he may produce a considerable rise or fall of potential, and greatly affect the flow of the current through the primary. In the experiment before shown, in which a brush appears at a wire attached to one terminal, and the wire is vibrated when the experimenter brings his insulated body in contact with the other terminal of the coil, the sudden rise of potential was made evident.

I may show you the behavior of the coil in another manner which possesses a feature of some interest. I have here a little light fan of aluminum sheet, fastened to a needle and arranged to rotate freely in a metal piece screwed to one of the terminals of the coil. When the coil is set to work, the molecules of the air are rhythmically attracted and repelled. As the force with which they are repelled is greater than that with which they are attracted, it results that there is a repulsion exerted on the surfaces of the fan. If the fan were made simply of a metal sheet, the repulsion would be equal on the opposite sides, and would produce no effect. But if one of the opposing surfaces is screened, or if, generally speaking, the bombardment on this side is weakened in some way or other, there remains the repulsion exerted upon the other, and the fan is set in rotation. The screening is best affected by fastening upon one of the opposing sides of the fan insulated conducting coatings, or, if the fan is made in the shape of an ordinary propeller screw, by fastening on one side, and close to it, an insulated metal plate. The static screen may, however, be omitted, and simply a thickness of insulating material fastened to one of the sides of the fan.

To show the behavior of the coil, the fan may be placed upon the terminal, and it will readily rotate when the coil is operated by currents of very high frequency. With a steady potential, of course, and even with alternating currents of very low frequency, it would not turn, because of the very slow exchange of air and, consequently, smaller bombardment; but in the latter case it might turn if the potential were excessive. With a pin wheel, quite the opposite rule holds good; it rotates best with a steady potential, and the effort is the smaller the higher the frequency. Now, it is very easy to adjust the conditions so that the potential is normally not sufficient to turn the fan, but that by connecting the other terminal of the coil with an insulated body it rises to a much greater value, so as to rotate the fan, and it is likewise possible to stop the rotation by connecting to the terminal a body of different size, thereby diminishing the potential.

Instead of using the fan in this experiment, we may use the "electric" radiometer with similar effect. But in this case, it will be found that the vanes will rotate only at high exhaustion or at ordinary pressures; they will not rotate at moderate pressures when the air is highly conducting. This curious observation was made conjointly by Professor Crookes and myself. I attribute the result to the high conductivity of the air, the molecules of which then do not act as independent carriers of electric charges, but act all together as a single conducting body. In such case, of course, if there is any repulsion at all of the molecules from the vanes, it must be very small. It is possible, however, that the result is in part due to the fact that the greater part of the discharge passes from the leading-in wire through the highly conducting gas, instead of passing off from the conducting vanes.

In trying the preceding experiment with the electric radiometer, the potential should not exceed a certain limit, as then the electrostatic attraction between the vanes and the glass of the bulb may be so great as to stop the rotation.

A most curious feature of alternate currents of high frequencies and potentials is that they enable us to perform many experiments by the use of one wire only. In many respects this feature is of great interest.

In a type of alternate current motor invented by me some years ago I produced rotation by inducing, by means of a single alternating current passed through a motor circuit, in the mass or other circuits of the motor, secondary currents, which, jointly with the primary or inducing current, created a moving field of force. A simple but crude form of such a motor is obtained by winding upon an iron core a primary, and close to it a secondary coil, joining the ends of the latter and placing a freely movable metal disc within the influence of the field produced by both. The iron core is employed for obvious reasons, but it is not essential to the operation. To improve the motor, the iron core is made to encircle the armature. Again, to improve, the secondary coil is made to overlap partly the primary, so that it cannot free itself from a strong inductive action of the latter, repel its lines as it may. Once more to improve, the proper difference of phase is obtained between the primary and secondary currents by a condenser, self-induction, resistance, or equivalent windings.

I had discovered, however, that rotation is produced by means of a single coil and core; my explanation of the phenomenon, and leading thought in trying the experiment, being that there must be a true time lag in the magnetization of the core. I remember the pleasure I had when, in the writings of Professor Ayrton, which came later to my hand, I found the idea of the time lag advocated. Whether there is a true time lag, or whether the retardation is due to eddy currents circulating in minute paths, must remain an open question, but the fact is that a coil wound upon an iron core and traversed by an alternating current creates a moving field of force, capable of setting an armature in rotation. It is of some interest, in conjunction with the historical Arago experiment, to mention that in lag or phase motors I have produced rotation in the opposite direction to the moving field, which means that in that experiment the magnet may not rotate, or may even rotate in the opposite direction to the moving disc. Here, then, is a motor (diagrammatically illustrated in Fig. 17), comprising a coil and iron core, and a freely movable copper disc in proximity to the latter.

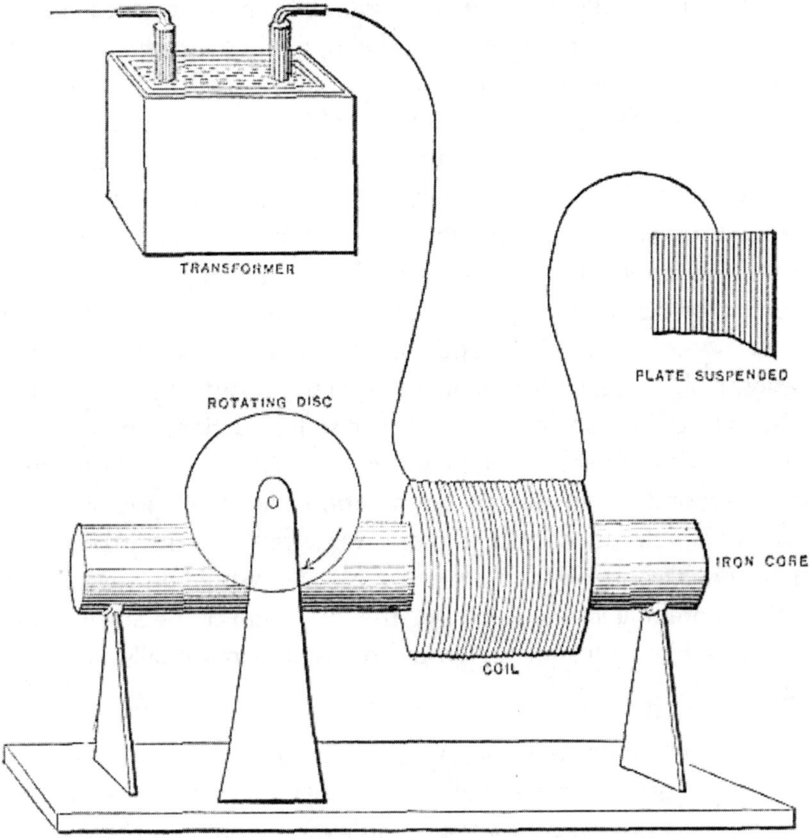


FIG. 17.—SINGLE WIRE AND "NO-WIRE" MOTOR.

To demonstrate a novel and interesting feature, I have, for a reason which I will explain, selected this type of motor. When the ends of the coil are connected to the terminals of an alternator the disc is set in rotation. But it is not this experiment, now well known, which I desire to perform. What I wish to show you is that this motor rotates with *one single* connection between it and the generator; that is to say, one terminal of the motor is connected to one terminal of the generator in this case the secondary of a high-tension induction coil, the other terminals of motor and generator being insulated in space. To produce rotation, it is generally (but not absolutely) necessary to connect the free end of the motor coil to an insulated body of some size. The experimenter's body is more than sufficient. If he touches the free terminal with an object held in the hand, a current pass through the coil and the copper disc is set in rotation. If an exhausted tube is put in series with the coil, the tube lights brilliantly, showing the passage of a strong current. Instead of the experimenter's body, a small metal sheet suspended on a cord may be used with the same result. In this case the plate acts as a condenser in series with the coil. It counteracts the self-induction of the latter and allows a strong current to pass. In such a combination, the greater the self-induction of the coil the smaller need be the plate, and this means that a lower frequency, or eventually a lower potential, is required to operate the motor. A single coil wound upon a core has a high self-induction; for this reason, principally, this type of motor was chosen to perform the experiment. Were a secondary closed coil wound upon the core, it would tend to diminish the self-induction, and then it would be necessary to employ a much higher frequency and potential. Neither would be advisable, for a higher potential would endanger the insulation of the small primary coil, and a higher frequency would result in a materially diminished torque.

It should be remarked that when such a motor with a closed secondary is used, it is not at all easy to obtain rotation with excessive frequencies, as the secondary cuts off almost completely the lines of the primary and this, of course, the more, the higher the frequency and allows the passage of but a minute current. In such a case, unless the secondary is closed through a condenser, it is almost essential, in order to produce rotation, to make the primary and secondary coils overlap each other more or less.

But there is an additional feature of interest about this motor, namely, it is not necessary to have even a single connection between the motor and generator, except, perhaps, through the ground: for not only is an insulated plate capable of giving off energy into space, but it is likewise capable of deriving it from an alternating electrostatic field, though in the latter case the available energy is much smaller. In this instance one of the motor terminals is connected to the insulated plate or body located within the alternating electrostatic field, and the other terminal preferably to the ground.

It is quite possible, however, that such "no-wire" motors, as they might be called, could be operated by conduction through the rarefied air at considerable distances. Alternate currents, especially of high frequencies, pass with astonishing freedom through even slightly rarefied gases. The upper strata of the air are rarefied. To reach a number of miles out into space requires the overcoming of difficulties of a merely mechanical nature. There is no doubt that with the enormous potentials obtainable by the use of high frequencies and oil insulation luminous discharges might be passed through many miles of rarefied air, and that, by thus directing the energy of many hundreds or thousands of horsepower, motors or lamps might be operated at considerable distances from stationary sources. But such schemes are mentioned merely as possibilities. We shall have no need to *transmit* power in this way. We shall have no need to *transmit* power at all. Ere many generations pass, our machinery will be driven by a power obtainable at any point of the universe. This idea is not novel. Men have been led to it long ago by instinct or reason. It has been expressed in many ways, and in many places, in the history of old and new. We find it in the delightful myth of Antheus, who derives power from the earth; we find it among the subtitle speculations of one of your splendid mathematicians, and in many hints and statements of thinkers of the present time. Throughout space there is energy. Is this energy static or kinetic? If static our hopes are in vain; if kinetic and this we know it is, for certain then it is a mere question of time when men will succeed in attaching their machinery to the very wheelwork of nature. Of all, living or dead, Crookes came nearest to doing it. His radiometer will turn in the light of day and in the darkness of the night; it will turn everywhere where there is heat, and heat is everywhere.

But, unfortunately, this beautiful little machine, while it goes down to posterity as the most interesting, must likewise be put on record as the most inefficient machine ever invented!

The preceding experiment is only one of many equally interesting experiments which may be performed by the use of only one wire with alternate currents of high potential and frequency. We may connect an insulated line to a source of such currents, we may pass an inappreciable current over the line, and on any point of the same we are able to obtain a heavy current, capable of fusing a thick copper wire. Or we may, by the help of some artifice, decompose a solution in any electrolytic cell by connecting only one pole of the cell to the line or source of energy. Or we may, by attaching to the line, or only bringing into its vicinity, light up an incandescent lamp, an exhausted tube, or a phosphorescent bulb.

However impracticable this plan of working may appear in many cases, it certainly seems practicable, and even recommendable, in the production of light. A perfected lamp would require but little energy, and if wires were used at all we ought to be able to supply that energy without a return wire.

It is now a fact that a body may be rendered incandescent or phosphorescent by bringing it either in single contact or merely in the vicinity of a source of electric impulses of the proper character, and that in this manner a quantity of light sufficient to afford a practical illuminant may be produced. It is, therefore, to say the least, worthwhile to attempt to determine the best conditions and to invent the best appliances for attaining this object.

Some experiences have already been gained in this direction, and I will dwell on them briefly, in the hope that they might prove useful.

The heating of a conducting body enclosed in a bulb and connected to a source of rapidly alternating electric impulses, is dependent on so many things of a different nature, that it would be difficult to give a generally applicable rule under which the maximum heating occurs.

As regards the size of the vessel, I have lately found that at ordinary or only slightly differing atmospheric pressures, when air is a good insulator, and hence practically the same amount of energy by a certain potential and frequency is given off from the body, whether the bulb be small or large, the body is brought to a higher temperature if enclosed in a small bulb, because of the better confinement of heat in this case.

At lower pressures, when air becomes more or less conducting, or if the air be sufficiently warmed as to become conducting, the body is rendered more intensely incandescent in a large bulb, obviously because, under otherwise equal conditions of test, more energy may be given off from the body when the bulb is large.

At very high degrees of exhaustion, when the matter in the bulb becomes "radiant," a large bulb has still an advantage, but a comparatively slight one, over the small bulb.

Finally, at excessively high degrees of exhaustion, which cannot be reached except by the employment of special means, there seems to be, beyond a certain and rather small size of vessel, no perceptible difference in the heating.

These observations were the result of a number of experiments, of which one, showing the effect of the size of the bulb at a high degree of exhaustion, may be described and shown here, as it presents a feature of interest. Three spherical bulbs of 2 inches, 3 inches and 4 inches diameter were taken, and in the center of each was mounted an equal length of an ordinary incandescent lamp filament of uniform thickness. In each bulb the piece of filament was fastened to the leading-in wire of platinum, contained in a glass stem sealed in the bulb; care being taken, of course, to make everything as nearly alike as possible. On each glass stem in the inside of the bulb was slipped a highly polished tube made of aluminum sheet, which fitted the stem and was held on it by spring pressure. The function of this aluminum tube will be explained subsequently. In each bulb an equal length of filament protruded above the metal tube. It is sufficient to say now that under these conditions' equal lengths of filament of the same thickness in other words, bodies of equal bulk were brought to incandescence. The three bulbs were sealed to a glass tube, which was connected to a Sprengel pump. When a high vacuum had been reached, the glass tube carrying the bulbs was sealed off. A current was then turned on successively on each bulb, and it was found that the filaments came to about the same brightness, and, if anything, the smallest bulb, which was placed midway between the two larger ones, may have been slightly brighter. This result was expected, for when either of the bulbs was connected to the coil the luminosity spread through the other two, hence the three bulbs constituted really one vessel. When all the three bulbs were connected in multiple arcs to the coil, in the largest of them the filament glowed brightest, in the next smaller it was a little less bright, and in the smallest, it only came to redness. The bulbs were then sealed off and separately tried. The brightness of the filaments was now such as would have been expected on the supposition that the energy given off was proportionate to the surface of the bulb, this surface in each case representing one of the coatings of a condenser. Accordingly, time was less difference between the largest and the middle sized than between the latter and the smallest bulb.

An interesting observation was made in this experiment. The three bulbs were suspended from a straight bare wire connected to a terminal of the coil, the largest bulb being placed at the end of the wire, at some distance from it the smallest bulb, and an equal distance from the latter the middle-sized one. The carbons glowed then in both the larger bulbs about as expected, but the smallest did not get its share by far. This observation led me to exchange the position of the bulbs, and I then observed that whichever of the bulbs was in the middle it was by far less bright than it was in any other position. This mystifying result was, of course, found to be due to the electrostatic action between the bulbs. When they were placed at a considerable distance, or when they were attached to the corners of an equilateral triangle of copper wire, they glowed about in the order determined by their surfaces.

As to the shape of the vessel, it is also of some importance, especially at high degrees of exhaustion. Of all the possible constructions, it seems that a spherical globe with the refractory body mounted in its center is the best to employ. In experience it has been demonstrated that in such a globe a refractory body of a given bulk is more easily brought to incandescence than when otherwise shaped bulbs are used. There is also an advantage in giving to the incandescent body the shape of a sphere, for self-evident reasons. In any case the body should be mounted in the center, where the atoms rebounding from the glass collide. This object is best attained in the spherical bulb; but it is also attained in a cylindrical vessel with one or two straight filaments coinciding with its axis, and possibly also in parabolical or spherical bulbs with the refractory body or bodies placed in the focus or foci of the same; though the latter is not probable, as the electrified atoms should in all cases rebound normally from the surface they strike, unless the speed were excessive, in which case they *would* probably follow the general law of reflection. No matter what shape the vessel may have, if the exhaustion be low, a filament mounted in the globe is brought to the same degree of incandescence in all parts; but if the exhaustion be high and the bulb be spherical or pear-shaped, as usual, focal points form and the filament is heated to a higher degree at or near such points.

To illustrate the effect, I have here two small bulbs which are alike, only one is exhausted to a low and the other to a very high degree. When connected to the coil, the filament in the former glows uniformly throughout all its length, whereas in the latter, that portion of the filament which is in the center of the bulb glows far more intensely than the rest. A curious point is that the phenomenon occurs even if two filaments are mounted in a bulb, each being connected to one terminal of the coil, and, what is still more curious, if they be very near together, provided the vacuum be very high. I noted in experiments with such bulbs that the filaments would give way usually at a certain point, and in the first trials I attributed it to a defect in the carbon. But when the phenomenon occurred many times in succession, I recognized its real cause.

In order to bring a refractory body enclosed in a bulb to incandescence, it is desirable, on account of economy, that all the energy supplied to the bulb from the source should reach without loss the body to be heated; from there, and from nowhere else, it should be radiated. It is, of course, out of the question to reach this theoretical result, but it is possible by a proper construction of the illuminating device to approximate it more or less.

For many reasons, the refractory body is placed in the center of the bulb, and it is usually supported on a glass stem containing the leading-in wire. As the potential of this wire is alternated, the rarefied gas surrounding the stem is acted upon inductively, and the glass stem is violently bombarded and heated. In this manner by far the greater portion of the energy supplied to the bulb especially when exceedingly high frequencies are used may be lost for the purpose contemplated. To obviate this loss, or at least to reduce it to a minimum, I usually screen the rarefied gas surrounding the stem from the inductive action of the leading-in wire by providing the stem with a tube or coating of conducting material. It seems beyond doubt that the best among metals to employ for this purpose is aluminum, on account of its many remarkable properties. Its only fault is that it is easily fusible, and, therefore, its distance from the incandescing body should be properly estimated. Usually, a thin tube, of a diameter somewhat smaller than that of the glass stem, is made of the finest aluminum sheet, and slipped on the stem. The tube is conveniently prepared by wrapping around a rod fastened in a lathe a piece of aluminum sheet of the proper size, grasping the sheet firmly with clean chamois leather or blotting paper, and spinning the rod very fast. The sheet is wound tightly around the rod, and a highly polished tube of one or three layers of the sheet is obtained. When slipped on the stem, the pressure is generally sufficient to prevent it from slipping off, but, for safety, the lower edge of the sheet may be turned inside. The upper inside corner of the sheet that is, the one which is nearest to the refractory incandescent body should be cut out diagonally, as it often happens that, in consequence of the intense heat, this corner turns toward the inside and comes very near to, or in contact with, the wire, or filament, supporting the refractory body. The greater part of the energy supplied to the bulb is then used up in heating the metal tube, and the bulb is rendered useless for the purpose. The aluminum sheet should project above the glass stem more or less one inch or so, or else, if the glass be too close to the incandescing body, it may be strongly heated and become more or less conducting, whereupon it may be ruptured, or may, by its conductivity, establish a good electrical connection between the metal tube and the leading-in wire, in which case, again, most of the energy will be lost in heating the former. Perhaps the best way is to make the top of the glass tube, for about an inch, of a much smaller diameter. To still further reduce the danger arising from the heating of the glass stem, and also with the view of preventing an electrical connection between the metal tube and the electrode, I preferably wrap the stem with several layers of thin mica, which extends at least as far as the metal tube. In some bulbs I have also used an outside insulating cover.

The preceding remarks are only made to aid the experimenter in the first trials, for the difficulties which he encounters he may soon find means to overcome in his own way.

To illustrate the effect of the screen, and the advantage of using it, I have here two bulbs of the same size, with their stems, leading-in wires and incandescent lamp filaments tied to the latter, as nearly alike as possible. The stem of one bulb is provided with an aluminum tube, the stem of the other has none. Originally the two bulbs were joined by a tube which was connected to a Sprengel pump. When a high vacuum had been reached, first the connecting tube, and then the bulbs, were sealed off; they are therefore of the same degree of exhaustion. When they are separately connected to the coil giving a certain potential, the carbon filament in the bulb provided with the aluminum screen is rendered highly incandescent, while the filament in the other bulb may, with the same potential, not even come to redness, although in reality the latter bulb takes generally more energy than the former. When they are both connected together to the terminal, the difference is even more apparent, showing the importance of the screening. The metal tube placed on the stem containing the leading-in wire performs really two distinct functions: First: it acts more or less as an electrostatic screen, thus economizing the energy supplied to the bulb; and, second, to whatever extent it may fail to act electrostatically, it acts mechanically, preventing the bombardment, and consequently intense heating and possible deterioration of the slender support of the refractory incandescent body, or of the glass stem containing the leading-in wire. I say *slender* support, for it is evident that in order to confine the heat more completely to the incandescing body its support should be very thin, so as to carry away the smallest possible amount of heat by conduction. Of all the supports used I have found an ordinary incandescent lamp filament to be the best, principally because among conductors it can withstand the highest degrees of heat.

The effectiveness of the metal tube as an electrostatic screen depends largely on the degree of exhaustion.

At excessively high degrees of exhaustion which are reached by using great care and special means in connection with the Sprengel pump when the matter in the globe is in the ultra-radiant state, it acts most perfectly. The shadow of the upper edge of the tube is then sharply defined upon the bulb.

At a somewhat lower degree of exhaustion, which is about the ordinary "non-striking" vacuum, and generally as long as the matter moves predominantly in straight lines, the screen still does well. In elucidation of the preceding remark, it is necessary to state that what is a "non-striking" vacuum for a coil operated, as ordinarily, by impulses, or currents, of low frequency, is not, by far, so when the coil is operated by currents of very high frequency. In such case the discharge may pass with great freedom through the rarefied gas through which a low-frequency discharge may not pass, even though the potential be much higher. At ordinary atmospheric pressures just, the reverse rule holds good: the higher the frequency, the less the spark discharge is able to jump between the terminals, especially if they are knobs or spheres of some size.

Finally, at very low degrees of exhaustion, when the gas is well conducting, the metal tube not only does not act as an electrostatic screen, but even is a drawback, aiding to a considerable extent the dissipation of the energy laterally from the leading-in wire. This, of course, is to be expected. In this case, namely, the metal tube is in good electrical connection with the leading-in wire, and most of the bombardment is directed upon the tube. As long as the electrical connection is not good, the conducting tube is always of some advantage, for although it may not greatly economize energy, still it protects the support of the refractory button, and is a means for concentrating more energy upon the same.

To whatever extent the aluminum tube performs the function of a screen, its usefulness is therefore limited to very high degrees of exhaustion when it is insulated from the electrode that is, when the gas as a whole is non-conducting, and the molecules, or atoms, act as independent carriers of electric charges.

In addition to acting as a more or less effective screen, in the true meaning of the word, the conducting tube or coating may also act, by reason of its conductivity, as a sort of equalizer or dampener of the bombardment against the stem. To be explicit, I assume the action as follows: Suppose a rhythmical bombardment to occur against the conducting tube by reason of its imperfect action as a screen, it certainly must happen that some molecules, or atoms, strike the tube sooner than others. Those which come first in contact with it give up their superfluous charge, and the tube is electrified, the electrification instantly spreading over its surface. But this must diminish the energy lost in the bombardment for two reasons: first, the charge given up by the atoms spreads over a great area, and hence the electric density at any point is small, and the atoms are repelled with less energy than they would be if they would strike against a good insulator: secondly, as the tube is electrified by the atoms which first come in contact with it, the progress of the following atoms against the tube is more or less checked by the repulsion which the electrified tube must exert upon the similarly electrified atoms. This repulsion may perhaps be sufficient to prevent a large portion of the atoms from striking the tube, but at any rate it must diminish the energy of their impact. It is clear that when the exhaustion is very low, and the rarefied gas well conducting, neither of the above effects can occur, and, on the other hand, the fewer the atoms, with the greater freedom they move; in other words, the higher the degree of exhaustion, up to a limit, the more telling will be both the effects.

What I have just said may afford an explanation of the phenomenon observed by Prof. Crookes, namely, that a discharge through a bulb is established with much greater facility when an insulator than when a conductor is present in the same. In my opinion, the conductor acts as a dampener of the motion of the atoms in the two ways pointed out; hence, to cause a visible discharge to pass through the bulb, a much higher potential is needed if a conductor, especially of much surface, be present.

For the sake of clearness of some of the remarks before made, I must now refer to Figs. 18, 19 and 20, which illustrate various arrangements with a type of bulb most generally used.

Fig. 18 is a section through a spherical bulb L, with the glass stem s, containing the leading-in wire w; which has a lamp filament l fastened to it, serving to support the refractory button m in the center. M is a sheet of thin mica wound in several layers around the stem s, and a is the aluminum tube.

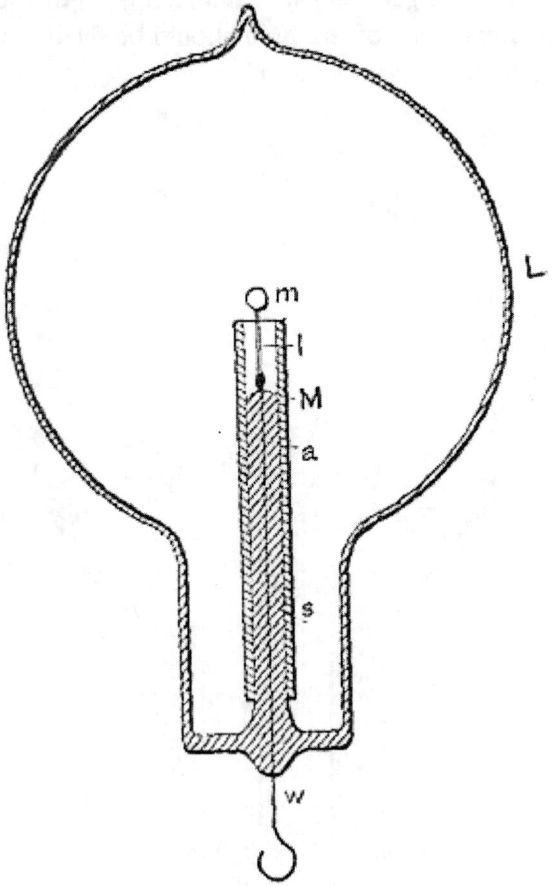


FIG. 18.—BULB WITH MICA TUBE AND ALUMINIUM SCREEN.

Fig. 19 illustrates such a bulb in a somewhat more advanced stage of perfection. A metallic tube S is fastened by means of some cement to the neck of the tube. In the tube is screwed a plug P, of insulating material, in the center of which is fastened a metallic terminal t, for the connection to the leading-in wire w. This terminal must be well insulated from the metal tube S, therefore, if the cement used is conducting and most generally it is sufficiently so the space between the plug P and the neck of the bulb should be filled with some good insulating material, as mica powder.

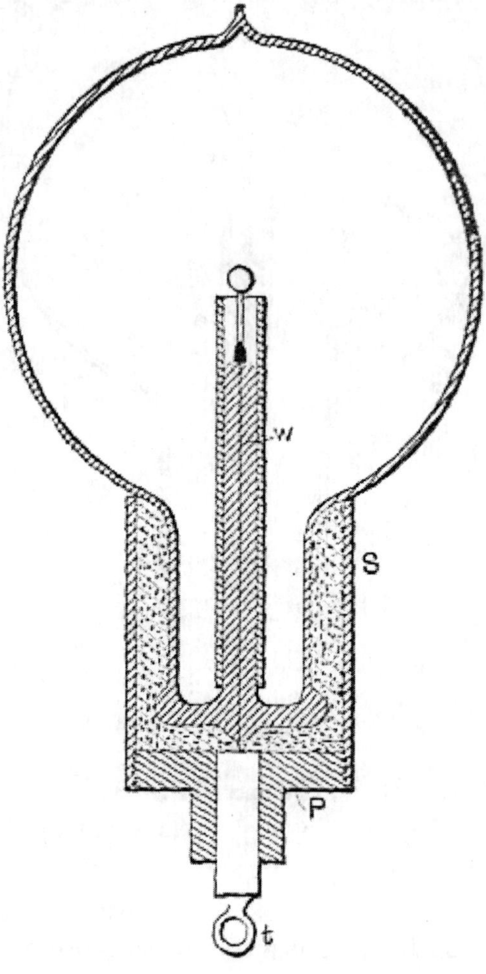


FIG. 19.—IMPROVED BULB WITH SOCKET AND SCREEN.

Fig. 20 shows a bulb made for experimental purposes. In this bulb the aluminum tube is provided with an external connection, which serves to investigate the effect of the tube under various conditions. It is referred to chiefly to suggest a line of experiment followed.

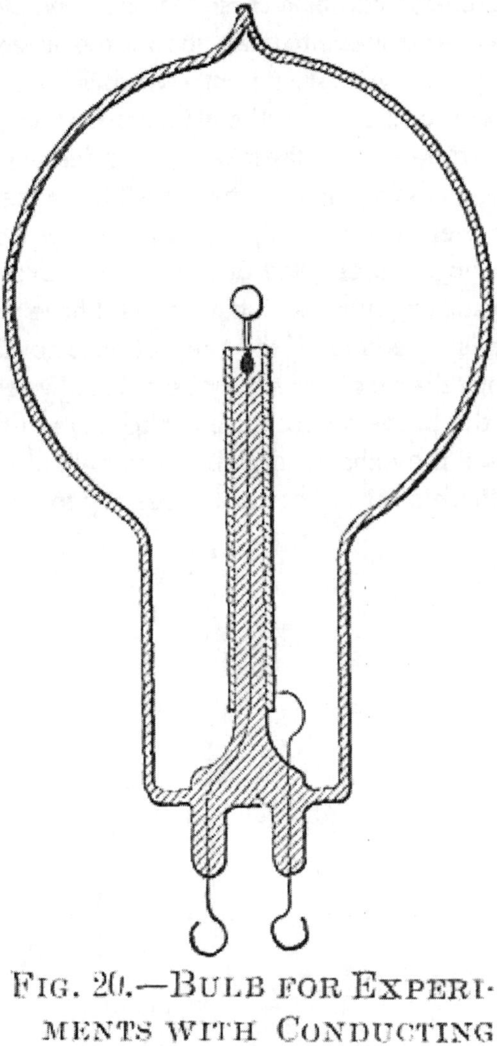

FIG. 20.—BULB FOR EXPERIMENTS WITH CONDUCTING TUBE.

Since the bombardment against the stem containing the leading-in wire is due to the inductive action of the latter upon the rarefied gas, it is of advantage to reduce this action as far as practicable by employing a very thin wire, surrounded by a very thick insulation of glass or other material, and by making the wire passing through the rarefied gas as short as practicable. To combine these features, I employ a large tube T (Fig. 21), which protrudes into the bulb to some distance, and carries on the top a very short glass stem s, into which is sealed the leading-in wire w, and I protect the top of the glass stem against the heat by a small, aluminum tube a and a layer of mica underneath the same, as usual. The wire w, passing through the large tube to the outside of the bulb, should be well insulated with a glass tube, for instance and the space between ought to be filled out with some excellent insulator. Among many insulating powders I have tried, I have found that mica powder is the best to employ. If this precaution is not taken, the tube T, protruding into the bulb, will surely be cracked in consequence of the heating by the brushes which are apt to form in the upper part of the tube, near the exhausted globe, especially if the vacuum be excellent, and therefore the potential necessary to operate the lamp very high.

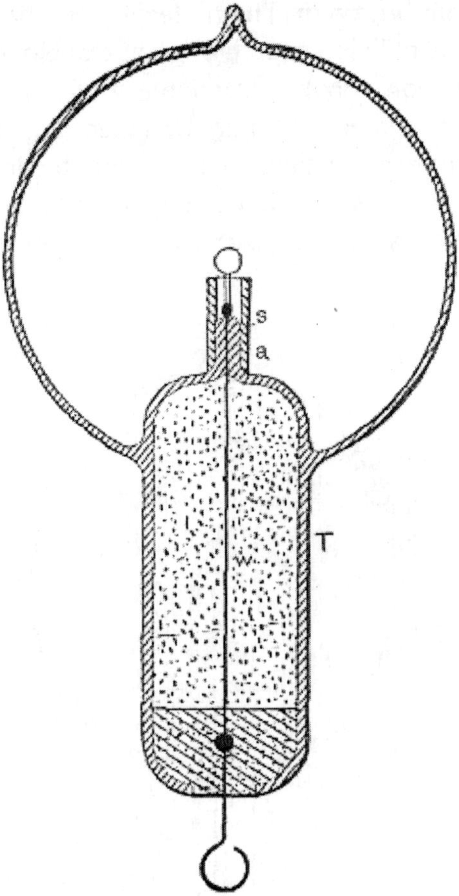


FIG. 21.—IMPROVED BULB WITH NON-CONDUCTING BUTTON.

Fig. 22 illustrates a similar arrangement, with a large tube T protruding into the part of the bulb containing the refractors button m. In this case the wire leading from the outside into the bulb is omitted, the energy required being supplied through condenser coatings C C. The insulating packing P should in this construction be tightly fitting to the glass, and rather wide, or otherwise the discharge might avoid passing through the wire w, which connects the inside condenser coating to the incandescent button m. The molecular bombardment against the glass stem in the bulb is a source of great trouble. As illustration I will cite a phenomenon only too frequently and unwillingly observed. A bulb, preferably a large one, may be taken, and a good conducting body, such as a piece of carbon, may be mounted in it upon a platinum wire sealed in the glass stem. The bulb may be exhausted to a fairly high degree, nearly to the point when phosphorescence begins to appear.

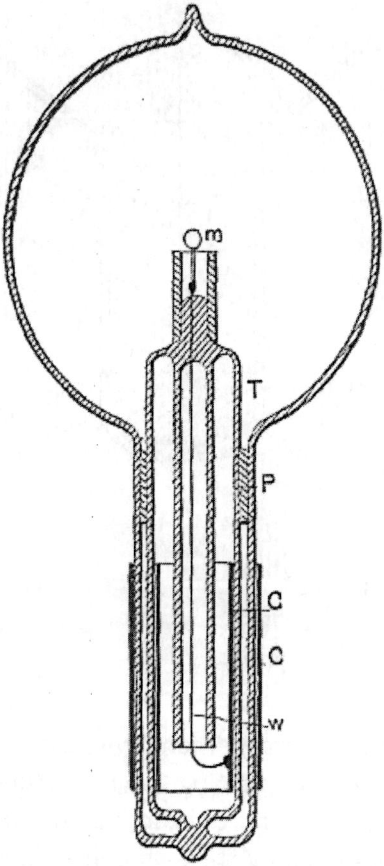


FIG. 22.—TYPE OF BULB WITHOUT LEADING-IN WIRE.

When the bulb is connected with the coil, the piece of carbon, if small, may become highly incandescent at first, but its brightness immediately diminishes, and then the discharge may break through the glass somewhere in the middle of the stem, in the form of bright sparks, in spite of the fact that the platinum wire is in good electrical connection with the rarefied gas through the piece of carbon or metal at the top. The first sparks are singularly bright, recalling those drawn from a clear surface of mercury. But, as they heat the glass rapidly, they, of course, lose their brightness, and cease when the glass at the ruptured place becomes incandescent, or generally sufficiently hot to conduct. When observed for the first time the phenomenon must appear very curious and shows in a striking manner how radically different alternate currents, or impulses, of high frequency behave, as compared with steady currents, or currents of low frequency. With such currents namely, the latter the phenomenon would of course not occur. When frequencies such as are obtained by mechanical means are used, I think that the rupture of the glass is more or less the consequence of the bombardment, which warms it up and impairs its insulating power; but with frequencies obtainable with condensers, I have no doubt that the glass may give way without previous heating. Although this appears most singular at first, it is in reality what we might expect to occur. The energy supplied to the wire leading into the bulb is given off partly by direct action through the carbon button, and partly by inductive action through the glass surrounding the wire. The case is thus analogous to that in which a condenser shunted by a conductor of low resistance is connected to a source of alternating currents. As long as the frequencies are low, the conductor gets the most, and the condenser is perfectly safe: but when the frequency becomes excessive, the *role* of the conductor may become quite insignificant. In the latter case the difference of potential at the terminals of the condenser may become so great as to rupture the dielectric, notwithstanding the fact that the terminals are joined by a conductor of low resistance.

It is, of course, not necessary, when it is desired to produce the incandescence of a body enclosed in a bulb by means of these currents, that the body should be a conductor, for even a perfect non-conductor may be quite as readily heated. For this purpose, it is sufficient to surround a conducting electrode with a non-conducting material, as, for instance, in the bulb described before in Fig. 21, in which a thin incandescent lamp filament is coated with a non-conductor, and supports a button of the same material on the top. At the start the bombardment goes on by inductive action through the non-conductor, until the same is sufficiently heated to become conducting, when the bombardment continues in the ordinary way.

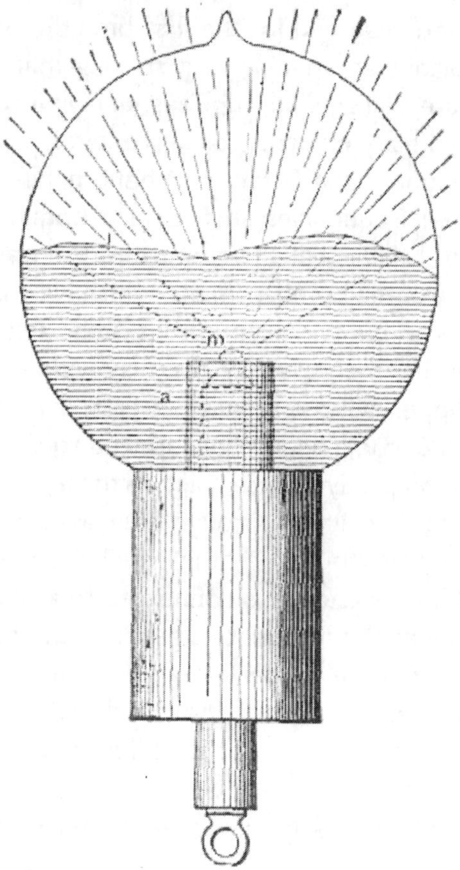


FIG. 23.—EFFECT PRODUCED BY A RUBY DROP

A different arrangement used in some of the bulbs constructed is illustrated in Fig. 23. In this instance a non-conductor m is mounted in a piece of common arc light carbon so as to project some small distance above the latter. The carbon piece is connected to the leading-in wire passing through a glass stem, which is wrapped with several layers of mica. An aluminum tube a is employed as usual for screening. It is so arranged that it reaches very nearly as high as the carbon and only the non-conductor m projects a little above it. The bombardment goes at first against the upper surface of carbon, the lower parts being protected by the aluminum tube. As soon, however, as the non-conductor m is heated it is rendered good conducting, and then it becomes the center of the bombardment, being most exposed to the same.

I have also constructed during these experiments many such single-wire bulbs with or without internal electrode, in which the radiant matter was projected against, or focused upon, the body to be rendered incandescent. Fig. 24 illustrates one of the bulbs used. It consists of a spherical globe L, provided with a long neck n, on the top, for increasing the action in some cases by the application of an external conducting coating. The globe L is blown out on the bottom into a very small bulb b, which serves to hold it firmly in a socket S of insulating material into which it is cemented. A fine lamp filament f, supported on a wire w, passes through the center of the globe L. The filament is rendered incandescent in the middle portion, where the bombardment proceeding from the lower inside surface of the globe is most intense. The lower portion of the globe, as far as the socket S reaches, is rendered conducting, either by a tinfoil coating or otherwise, and the external electrode is connected to a terminal of the coil.

The arrangement diagrammatically indicated in Fig. 24 was found to be an inferior one when it was desired to render incandescent a filament or button supported in the center of the globe, but it was convenient when the object was to excite phosphorescence.

In many experiments in which bodies of a different kind were mounted in the bulb as, for instance, indicated in Fig. 23, some observations of interest were made.

It was found, among other things, that in such cases, no matter where the bombardment began, just as soon as a high temperature was reached there was generally one of the bodies which seemed to take most of the bombardment upon itself, the other, or others, being thereby relieved. This quality appeared to depend principally on the point of fusion, and on the facility with which the body was "evaporated," or, generally speaking, disintegrated meaning by the latter term not only the throwing off of atoms, but likewise of larger lumps. The observation made was in accordance with generally accepted notions. In a highly exhausted bulb electricity is carried off from the electrode by independent carriers, which are partly the atoms, or molecules, of the residual atmosphere, and partly the atoms, molecules, or lumps thrown off from the electrode. If the electrode is composed of bodies of different character, and if one of these is more easily disintegrated than the others, most of the electricity supplied is carried off from that body, which is then brought to a higher temperature than the others, and this the more, as upon an increase of the temperature the body is still more easily disintegrated.

It seems to me quite probable that a similar process takes place in the bulb even with a homogeneous electrode, and I think it to be the principal cause of the disintegration. There is bound to be some irregularity, even if the surface is highly polished, which, of course, is impossible with most of the refractory bodies employed as electrodes. Assume that a point of the electrode gets hotter, instantly most of the discharge passes through that point, and a minute patch is probably fused and evaporated. It is now possible that in consequence of the violent disintegration the spot attacked sinks in temperature, or that a counter force is created, as in an arc; at any rate, the local tearing off meets with the *limitations* incident to the experiment, whereupon the same process occurs on another place. To the eye the electrode appears uniformly brilliant, but there are upon it points constantly shifting and wandering around, of a temperature far above the mean, and this materially hastens the process of deterioration. That some such thing occurs, at least when the electrode is at a lower temperature, sufficient experimental evidence can be obtained in the following manner: Exhaust a bulb to a very high degree, so that with a fairly high potential the discharge cannot pass that is, not a luminous one, for a weak invisible discharge occurs always, in all probability. Now raise slowly and carefully the potential, leaving the primary current on no more than for an instant. At a certain point, two, three, or half a dozen phosphorescent spots will appear on the globe. These places of the glass are evidently more violently bombarded than others, this being due to the unevenly distributed electric density, necessitated, of course, by sharp projections, or, generally speaking, irregularities of the electrode. But the luminous patches are constantly changing in position, which is especially well observable if one manages to produce very few, and this indicates that the configuration of the electrode is rapidly changing.

From experiences of this kind, I am led to infer that, in order to be most durable, the refractory button in the bulb should be in the form of a sphere with a highly polished surface. Such a small sphere could be manufactured from a diamond or some other crystal, but a better way would be to fuse, by the employment of extreme degrees of temperature, some oxide as, for instance, zirconia into a small drop, and then keep it in the bulb at a temperature somewhat below its point of fusion.

Interesting and useful results can no doubt be reached in the direction of extreme degrees of heat. How can such high temperatures be arrived at? How are the highest degrees of heat reached in nature? By the impact of stars, by high speeds and collisions. In a collision any rate of heat generation may be attained. In a chemical process we are limited. When oxygen and hydrogen combine, they fall, metaphorically speaking, from a definite height. We cannot go very far with a blast, nor by confining heat in a furnace, but in an exhausted bulb we can concentrate any amount of energy upon a minute button. Leaving practicability out of consideration, this, then, would be the means which, in my opinion, would enable us to reach the highest temperature. But a great difficulty when proceeding in this way is encountered, namely, in most cases the body is carried off before it can fuse and form a drop. This difficulty exists principally with an oxide such as zirconia because it cannot be compressed in so hard a cake that it would not be carried off quickly. I endeavored repeatedly to fuse zirconia, placing it in a cup or arc light carbon as indicated in Fig. 23. It glowed with a most intense light, and the stream of the particles projected out of the carbon cup was of a vivid white: but whether it was compressed in a cake or made into a paste with carbon, it was carried off before it could be fused. The carbon cup containing the zirconia had to be mounted very low in the neck of a large bulb, as the heating of the glass by the projected particles of the oxide was so rapid that in the first trial the bulb was cracked almost in an instant when the current was turned on. The heating of the glass by the projected particles was found to be always greater when the carbon cup contained a body which was rapidly carried off, I presume because in such cases, with the same potential, higher speeds were reached, and also because, per unit of time, more matter was projected that is, more particles would strike the glass.

The before mentioned difficulty did not exist, however, when the body mounted in the carbon cup offered great resistance to deterioration. For instance, when an oxide was first fused in an oxygen blast and then mounted in the bulb, it melted very readily into a drop.

Generally during the process of fusion magnificent light effects were noted, of which it would be difficult to give an adequate idea. Fig. 23 is intended to illustrate the effect observed with a ruby drop. At first one may see a narrow funnel of white light projected against the top of the globe, where it produces an irregularly outlined phosphorescent patch. When the point of the ruby fuses the phosphorescence becomes very powerful; but as the atoms are projected with much greater speed from the surface of the drop, soon the glass gets hot and "tired," and now only the outer edge of the patch glows. In this manner an intensely phosphorescent, sharply defined line, l, corresponding to the outline of the drop, is produced, which spreads slowly over the globe as the drop gets larger. When the mass begins to boil, small bubbles and cavities are formed, which cause dark colored spots to sweep across the globe. The bulb may be turned downward without fear of the drop falling off, as the mass possesses considerable viscosity.

I may mention here another feature of some interest, which I believe to have noted in the course of these experiments, though the observations do not amount to a certitude. It appeared that under the molecular impact caused by the rapidly alternating potential the body was fused and maintained in that state at a lower temperature in a highly exhausted bulb than was the case at normal pressure and application of heat in the ordinary way that is, at least, judging from the quantity of the light emitted. One of the experiments performed may be mentioned here by way of illustration. A small piece of pumice stone was stuck on a platinum wire, and first melted to it in a gas burner. The wire was next placed between two pieces of charcoal and a burner applied so as to produce an intense heat, sufficient to melt down the pumice stone into a small glass-like button. The platinum wire had to be taken of sufficient thickness to prevent its melting in the fire. While in the charcoal fire, or when held in a burner to get a better idea of the degree of heat, the button glowed with great brilliancy. The wire with the button was then mounted in a bulb, and upon exhausting the same to a high degree, the current was turned on slowly so as to prevent the cracking of the button. The button was heated to the point of fusion, and when it melted it did not, apparently, glow with the same brilliancy as before, and this would indicate a lower temperature. Leaving out of consideration the observer's possible, and even probable, error, the question is, can a body under these conditions be brought from a solid to a liquid state with evolution of less light?

When the potential of a body is rapidly alternated it is certain that the structure is jarred. When the potential is very high, although the vibrations may be few say 20,000 per second, the effect upon the structure may be considerable. Suppose, for example, that a ruby is melted into a drop by a steady application of energy. When it forms a drop, it will emit visible and invisible waves, which will be in a definite ratio, and to the eye the drop will appear to be of a certain brilliancy. Next, suppose we diminish to any degree we choose the energy steadily supplied, and, instead, supply energy which rises and falls according to a certain law. Now, when the drop is formed, there will be emitted from it three different kinds of vibrations the ordinary visible, and two kinds of invisible waves: that is, the ordinary dark waves of all lengths, and, in addition, waves of a well-defined character. The latter would not exist by a steady supply of the energy; still, they help to jar and loosen the structure. If this really be the case, then the ruby drop will emit relatively less visible and more invisible waves than before. Thus, it would seem that when a platinum wire, for instance, is fused by currents alternating with extreme rapidity, it emits at the point of fusion less light and more invisible radiation than it does when melted by a steady current, though the total energy used up in the process of fusion is the same in both cases. Or, to cite another example, a lamp filament is not capable of withstanding as long with currents of extreme frequency as it does with steady currents, assuming that it be worked at the same luminous intensity. This means that for rapidly alternating currents the filament should be shorter and thicker. The higher the frequency that is, the greater the departure from the steady flow the worse it would be for the filament. But if the truth of this remark were demonstrated, it would be erroneous to conclude that such a refractory button as used in these bulbs would be deteriorated quicker by currents of extremely high frequency than by steady or low frequency currents. From experience I may say that just the opposite holds good: the button withstands the bombardment better with currents of very high frequency. But this is due to the fact that a high frequency discharge passes through a rarefied gas with much greater freedom than a steady or low frequency discharge, and this will say that with the former we can work with a lower potential or with a less violent impact. As long, then, as the gas is of no consequence, a steady or low frequency current is better; but as soon as the action of the gas is desired and important, high frequencies are preferable.

In the course of these experiments a great many trials were made with all kinds of carbon buttons. Electrodes made of ordinary carbon buttons were decidedly more durable when the buttons were obtained by the application of enormous pressure. Electrodes prepared by depositing carbon in well-known ways did not show up well; they blackened the globe very quickly. From many experiences I conclude that lamp filaments obtained in this manner can be advantageously used only with low potentials and low frequency currents. Some kinds of carbon withstand so well that, in order to bring them to the point of fusion, it is necessary to employ very small buttons. In this case the observation is rendered very difficult on account of the intense heat produced. Nevertheless, there can be no doubt that all kinds of carbon are fused under the molecular bombardment, but the liquid state must be one of great instability. Of all the bodies tried there were two which withstood best diamond and carborundum. These two showed up about equally, but the latter was preferable, for many reasons. As it is more than likely that this body is not yet generally known, I will venture to call your attention to it.

It has been recently produced by Mr. E.G. Acheson, of Monongahela City, Pa., U.S.A. It is intended to replace ordinary diamond powder for polishing precious stones, etc., and I have been informed that it accomplishes this object quite successfully. I do not know why the name "carborundum" has been given to it, unless there is something in the process of its manufacture which justifies this selection. Through the kindness of the inventor, I obtained a short while ago some samples which I desired to test in regard to their qualities of phosphorescence and capability of withstanding high degrees of heat.

Carborundum can be obtained in two forms in the form of "crystals" and of powder. The former appears to the naked eye dark colored but are very brilliant; the latter is of nearly the same color as ordinary diamond powder, but very much finer. When viewed under a microscope the samples of crystals given to me did not appear to have any definite form, but rather resembled pieces of broken up egg coal of fine quality. The majority were opaque, but there were some which were transparent and colored. The crystals are a kind of carbon containing some impurities; they are extremely hard and withstand for a long time even an oxygen blast. When the blast is directed against them, they at first form a cake of some compactness, probably in consequence of the fusion of impurities they contain. The mass withstands for a very long time the blast without further fusion; but a slow carrying off, or burning, occurs, and, finally, a small quantity of a glass-like residue is left, which, I suppose, is melted alumina. When compressed strongly they conduct very well, but not as well as ordinary carbon. The powder, which is obtained from the crystals in some way, is practically non-conducting. It affords a magnificent polishing material for stones.

The time has been too short to make a satisfactory study of the properties of this product, but enough experience has been gained in a few weeks I have experimented upon it to say that it does possess some remarkable properties in many respects. It withstands excessively high degrees of heat, it is little deteriorated by molecular bombardment, and it does not blacken the globe as ordinary carbon does. The only difficulty which I have found in its use in connection with these experiments was to find some binding material which would resist the heat and the effect of the bombardment as successfully as carborundum itself does.

I have here a number of bulbs which I have provided with buttons of carborundum. To make such a button of carborundum crystals I proceed in the following manner: I take an ordinary lamp filament and dip its point in tar, or some other thick substance or paint which may be readily carbonized. I next pass the point of the filament through the crystals, and then hold it vertically over a hot plate. The tar softens and forms a drop on the point of the filament, the crystals adhering to the surface of the drop. By regulating the distance from the plate, the tar is slowly dried out and the button becomes solid. I then once more dip the button in tar and hold it again over a plate until the tar is evaporated, leaving only a hard mass which firmly binds the crystals. When a larger button is required, I repeat the process several times, and I generally also cover the filament a certain distance below the button with crystals. The button being mounted in a bulb, when a good vacuum has been reached, first a weak and then a strong discharge is passed through the bulb to carbonize the tar and expel all gases, and later it is brought to a very intense incandescence.

When the powder is used, I have found it best to proceed as follows: I make a thick paint of carborundum and tar and pass a lamp filament through the paint. Taking then most of the paint off by rubbing the filament against a piece of chamois leather, I hold it over a hot plate until the tar evaporates and the coating becomes firm. I repeat this process as many times as it is necessary to obtain a certain thickness of coating. On the point of the coated filament, I form a button in the same manner.

There is no doubt that such a button properly prepared under great pressure of carborundum, especially of powder of the best quality, will withstand the effect of the bombardment fully as well as anything we know. The difficulty is that the binding material gives way, and the carborundum is slowly thrown off after some time. As it does not seem to blacken the globe in the least, it might be found useful for coating the filaments of ordinary incandescent lamps, and I think that it is even possible to produce thin threads or sticks of carborundum which will replace the ordinary filaments in an incandescent lamp. A carborundum coating seems to be more durable than other coatings, not only because the carborundum can withstand high degrees of heat, but also because it seems to unite with the carbon better than any other material I have tried. A coating of zirconia or any other oxide, for instance, is far more quickly destroyed. I prepared buttons of diamond dust in the same manner as of carborundum, and these came in durability nearest to those prepared of carborundum, but the binding paste gave way much more quickly in the diamond buttons: this, however, I attributed to the size and irregularity of the grains of the diamond.

It was of interest to find whether carborundum possesses the quality of phosphorescence. One is, of course, prepared to encounter two difficulties: first, as regards the rough product, the "crystals," they are good conducting, and it is a fact that conductors do not phosphoresce; second, the powder, being exceedingly fine, would not be apt to exhibit very prominently this quality, since we know that when crystals, even such as diamond or ruby, are finely powdered, they lose the property of phosphorescence to a considerable degree.

The question presents itself here, can a conductor phosphoresce? What is there in such a body as a metal, for instance, that would deprive it of the quality of phosphorescence, unless it is that property which characterizes it as a conductor? for it is a fact that most of the phosphorescent bodies lose that quality when they are sufficiently heated to become more or less conducting. Then, if a metal be in a large measure, or perhaps entirely, deprived of that property, it should be capable of phosphorescence. Therefore, it is quite possible that at some extremely high frequency, when behaving practically as a non-conductor, a metal or any other conductor might exhibit the quality of phosphorescence, even though it be entirely incapable of phosphorescing under the impact of a low-frequency discharge. There is, however, another possible way how a conductor might at least *appear* to phosphoresce.

Considerable doubt still exists as to what really is phosphorescence, and as to whether the various phenomena comprised under this head are due to the same causes. Suppose that in an exhausted bulb, under the molecular impact, the surface of a piece of metal or other conductor is rendered strongly luminous, but at the same time it is found that it remains comparatively cool, would not this luminosity be called phosphorescence? Now such a result, theoretically at least, is possible, for it is a mere question of potential or speed. Assume the potential of the electrode, and consequently the speed of the projected atoms, to be sufficiently high, the surface of the metal piece against which the atoms are projected would be rendered highly incandescent, since the process of heat generation would be incomparably faster than that of radiating or conducting away from the surface of the collision. In the eye of the observer a single impact of the atoms would cause an instantaneous flash, but if the impacts were repeated with sufficient rapidity, they would produce a continuous impression upon his retina. To him then the surface of the metal would appear continuously incandescent and of constant luminous intensity, while in reality the light would be either intermittent or at least changing periodically in intensity. The metal piece would rise in temperature until equilibrium was attained that is until the energy continuously radiated would equal that intermittently supplied. But the supplied energy might under such conditions not be sufficient to bring the body to any more than a very moderate mean temperature, especially if the frequency of the atomic impacts be very low just enough that the fluctuation of the intensity of the light emitted could not be detected by the eye. The body would now, owing to the manner in which the energy is supplied, emit a strong light, and yet be at a comparatively very low mean temperature. How could the observer call the luminosity thus produced? Even if the analysis of the light would teach him something definite, still he would probably rank it under the phenomena of phosphorescence. It is conceivable that in such a way both conducting and non-conducting bodies may be maintained at a certain luminous intensity, but the energy required would vary greatly vary with the nature and properties of the bodies.

These and some foregoing remarks of a speculative nature were made merely to bring out curious features of alternate currents or electric impulses. By their help we may cause a body to emit *more* light, while at a certain mean temperature, than it would emit if brought to that temperature by a steady supply; and, again, we may bring a body to the point of fusion and cause it to emit *less* light than when fused by the application of energy in ordinary ways. It all depends on how we supply the energy, and what kind of vibrations we set up: in one case the vibrations are more, in the other less, adapted to affect our sense of vision.

Some effects, which I had not observed before, obtained with carborundum in the first trials, I attributed to phosphorescence, but in subsequent experiments it appeared that it was devoid of that quality. The crystals possess a noteworthy feature. In a bulb provided with a single electrode in the shape of a small circular metal disc, for instance, at a certain degree of exhaustion the electrode is covered with a milky film, which is separated by a dark space from the glow filling the bulb. When the metal disc is covered with carborundum crystals, the film is far more intense, and snow-white. This I found later to be merely an effect of the bright surface of the crystals, for when an aluminum electrode was highly polished it exhibited more or less the same phenomenon. I made a number of experiments with the samples of crystals obtained, principally because it would have been of special interest to find that they are capable of phosphorescence, on account of their being conducting. I could not produce phosphorescence distinctly, but I must remark that a decisive opinion cannot be formed until other experimenters have gone over the same ground.

The powder behaved in some experiments as though it contained alumina, but it did not exhibit with sufficient distinctness the red of the latter. Its dead color brightens considerably under the molecular impact, but I am now convinced it does not phosphoresce. Still, the tests with the powder are not conclusive, because powdered carborundum probably does not behave like a phosphorescent sulphide, for example, which could be finely powdered without impairing the phosphorescence, but rather like powdered ruby or diamond, and therefore it would be necessary, in order to make a decisive test, to obtain it in a large lump and polish up the surface.

If the carborundum proves useful in connection with these and similar experiments, its chief value will be found in the production of coatings, thin conductors, buttons, or other electrodes capable of withstanding extremely high degrees of heat.

The production of a small electrode capable of withstanding enormous temperatures I regard as of the greatest importance in the manufacture of light. It would enable us to obtain, by means of currents of very high frequencies, certainly 20 times, if not more, the quantity of light which is obtained in the present incandescent lamp by the same expenditure of energy. This estimate may appear to many exaggerated, but in reality, I think it is far from being so. As this statement might be misunderstood, I think it necessary to clearly expose the problem with which in this line of work we are confronted, and the manner in which, in my opinion, a solution will be arrived at.

Anyone who begins a study of the problem will be apt to think that what is wanted in a lamp with an electrode is a very high degree of incandescence of the electrode. There he will be mistaken. The high incandescence of the button is a necessary evil, but what is really wanted is the high incandescence of the gas surrounding the button. In other words, the problem in such a lamp is to bring a mass of gas to the highest possible incandescence. The higher the incandescence, the quicker the mean vibration, the greater is the economy of the light production. But to maintain a mass of gas at a high degree of incandescence in a glass vessel, it will always be necessary to keep the incandescent mass away from the glass; that is, to confine it as much as possible to the central portion of the globe.

In one of the experiments this evening a brush was produced at the end of a wire. This brush was a flame, a source of heat and light. It did not emit much perceptible heat, nor did it glow with an intense light; but is it the less a flame because it does not scorch my hand? Is it the less a flame because it does not hurt my eye by its brilliancy? The problem is precisely to produce in the bulb such a flame, much smaller in size, but incomparably more powerful. Were there means at hand for producing electric impulses of a sufficiently high frequency, and for transmitting them, the bulb could be done away with, unless it were used to protect the electrode, or to economize the energy by confining the heat. But as such means are not at disposal, it becomes necessary to place the terminal in a bulb and rarefy the air in the same. This is done merely to enable the apparatus to perform the work which it is not capable of performing at ordinary air pressure. In the bulb we are able to intensify the action to any degree so far that the brush emits a powerful light. The intensity of the light emitted depends principally on the frequency and potential of the impulses, and on the electric density of the surface of the electrode. It is of the greatest importance to employ the smallest possible button, in order to push the density very far. Under the violent impact of the molecules of the gas surrounding it, the small electrode is of course brought to an extremely high temperature, but around it is a mass of highly incandescent gas, a flame photosphere, many hundred times the volume of the electrode. With a diamond, carborundum, or zirconia button the photosphere can be as much as one thousand times the volume of the button. Without much reflecting one would think that in pushing so far, the incandescence of the electrode it would be instantly volatilized. But after a careful consideration he would find that, theoretically, it should not occur, and in this fact which, however, is experimentally demonstrated lies principally the future value of such a lamp.

At first, when the bombardment begins, most of the work is performed on the surface of the button, but when a highly conducting photosphere is formed the button is comparatively relieved. The higher the incandescence of the photosphere the more it approaches in conductivity to that of the electrode, and the more, therefore, the solid and the gas form one conducting body. The consequence is that the further is forced the incandescence the more work, comparatively, is performed on the gas, and the less on the electrode. The formation of a powerful photosphere is consequently the very means for protecting the electrode. This protection, of course, is a relative one, and it should not be thought that by pushing the incandescence higher the electrode is actually less deteriorated. Still, theoretically, with extreme frequencies, this result must be reached, but probably at a temperature too high for most of the refractory bodies known. Given, then, an electrode which can withstand to a very high limit the effect of the bombardment and outward strain, it would be safe no matter how much it is forced beyond that limit. In an incandescent lamp quiet, different considerations apply. There the gas is not at all concerned: the whole of the work is performed on the filament; and the life of the lamp diminishes so rapidly with the increase of the degree of incandescence that economical reasons compel us to work it at a low incandescence. But if an incandescent lamp is operated with currents of very high frequency, the action of the gas cannot be neglected, and the rules for the most economical working must be considerably modified.

In order to bring such a lamp with one or two electrodes to a great perfection, it is necessary to employ impulses of very high frequency. The high frequency secures, among others, two chief advantages, which have a most important bearing upon the economy of the light production. First, the deterioration of the electrode is reduced by reason of the fact that we employ a great many small impacts, instead of a few violent ones, which quickly shatter the structure; secondly, the formation of a large photosphere is facilitated.

In order to reduce the deterioration of the electrode to the minimum, it is desirable that the vibration be harmonic, for any suddenness hastens the process of destruction. An electrode lasts much longer when kept at incandescence by currents, or impulses, obtained from a high-frequency alternator, which rise and fall more or less harmonically, than by impulses obtained from a disruptive discharge coil. In the latter case there is no doubt that most of the damage is done by the fundamental sudden discharges.

One of the elements of loss in such a lamp is the bombardment of the globe. As the potential is very high, the molecules are projected with great speed; they strike the glass, and usually excite a strong phosphorescence. The effect produced is very pretty, but for economical reasons it would be perhaps preferable to prevent, or at least reduce to the minimum, the bombardment against the globe, as in such case it is, as a rule, not the object to excite phosphorescence, and as some loss of energy results from the bombardment. This loss in the bulb is principally dependent on the potential of the impulses and on the electric density on the surface of the electrode. In employing very high frequencies the loss of energy by the bombardment is greatly reduced, for, first, the potential needed to perform a given amount of work is much smaller; and, secondly, by producing a highly conducting photosphere around the electrode, the same result is obtained as though the electrode were much larger, which is equivalent to a smaller electric density. But be it by the diminution of the maximum potential or of the density, the gain is affected in the same manner, namely, by avoiding violent shocks, which strain the glass much beyond its limit of elasticity. If the frequency could be brought high enough, the loss due to the imperfect elasticity of the glass would be entirely negligible. The loss due to bombardment of the globe may, however, be reduced by using two electrodes instead of one. In such case each of the electrodes may be connected to one of the terminals; or else, if it is preferable to use only one wire, one electrode may be connected to one terminal and the other to the ground or to an insulated body of some surface, as, for instance, a shade on the lamp. In the latter case, unless some judgment is used, one of the electrodes might glow more intensely than the other.

But on the whole, I find it preferable when using such high frequencies to employ only one electrode and one connecting wire. I am convinced that the illuminating device of the near future will not require for its operation more than one lead, and, at any rate, it will have no leading-in wire, since the energy required can be as well transmitted through the glass. In experimental bulbs the leading-in wire is most generally used on account of convenience, as in employing condenser coatings in the manner indicated in Fig. 22, for example, there is some difficulty in fitting the parts, but these difficulties would not exist if a great many bulbs were manufactured; otherwise the energy can be conveyed through the glass as well as through a wire, and with these high frequencies the losses are very small. Such illuminating devices will necessarily involve the use of very high potentials, and this, in the eyes of practical men, might be an objectionable feature. Yet, in reality, high potentials are not objectionable certainly not in the least as far as the safety of the devices is concerned.

There are two ways of rendering an electric appliance safe. One is to use low potentials; the other is to determine the dimensions of the apparatus so that it is safe no matter how high a potential is used. Of the two the latter seems to me the better way, for then the safety is absolute, unaffected by any possible combination of circumstances which might render even a low-potential appliance dangerous to life and property. But the practical conditions require not only the judicious determination of the dimensions of the apparatus; they likewise necessitate the employment of energy of the proper kind. It is easy, for instance, to construct a transformer capable of giving, when operated from an ordinary alternate current machine of low tension, say 50,000 volts, which might be required to light a highly exhausted phosphorescent tube, so that, in spite of the high potential, it is perfectly safe, the shock from it producing no inconvenience. Still, such a transformer would be expensive, and in itself inefficient; and, besides, what energy was obtained from it would not be economically used for the production of light. The economy demands the employment of energy in the form of extremely rapid vibrations. The problem of producing light has been likened to that of maintaining a certain high pitch note by means of a bell. It should be said a *barely audible* note; and even these words would not express it, so wonderful is the sensitiveness of the eye. We may deliver powerful blows at long intervals, waste a good deal of energy, and still not get what we want; or we may keep up the note by delivering frequent gentle taps, and get nearer to the object sought by the expenditure of much less energy. In the production of light, as far as the illuminating device is concerned, there can be only one rule that is, to use as high frequencies as can be obtained; but the means for the production and conveyance of impulses of such character impose, at present at least, great limitations. Once it is decided to use very high frequencies, the return wire becomes unnecessary, and all the appliances are simplified. By the use of obvious means the same result is obtained as though the return wire were used. It is sufficient for this purpose to bring in contact with the bulb, or merely in the vicinity of the same, an insulated body of some surface. The surface need, of course, be the smaller, the higher the frequency and potential used, and necessarily, also, the higher the economy of the lamp or other device.

This plan of working has been resorted to on several occasions this evening. So, for instance, when the incandescence of a button was produced by grasping the bulb with the hand, the body of the experimenter merely served to intensify the action. The bulb used was similar to that illustrated in Fig. 19, and the coil was excited to a small potential, not sufficient to bring the button to incandescence when the bulb was hanging from the wire; and incidentally, in order to perform the experiment in a more suitable manner, the button was taken so large that a perceptible time had to elapse before, upon grasping the bulb, it could be rendered incandescent. The contact with the bulb was, of course, quite unnecessary. It is easy, by using a rather large bulb with an exceedingly small electrode, to adjust the conditions so that the latter is brought to bright incandescence by the mere approach of the experimenter within a few feet of the bulb, and that the incandescence subsides upon his receding.

In another experiment, when phosphorescence was excited, a similar bulb was used. Here again, originally, the potential was not sufficient to excite phosphorescence until the action was intensified in this case, however, to present a different feature, by touching the socket with a metallic object held in the hand. The electrode in the bulb was a carbon button so large that it could not be brought to incandescence, and thereby spoil the effect produced by phosphorescence.

Again, in another of the early experiments, a bulb was used as illustrated in Fig. 12. In this instance, by touching the bulb with one or two fingers, one or two shadows of the stem inside were projected against the glass, the touch of the finger producing the same result as the application of an external negative electrode under ordinary circumstances.

In all these experiments the action was intensified by augmenting the capacity at the end of the lead connected to the terminal. As a rule, it is not necessary to resort to such means, and would be quite unnecessary with still higher frequencies; but when it is desired, the bulb, or tube, can be easily adapted to the purpose.

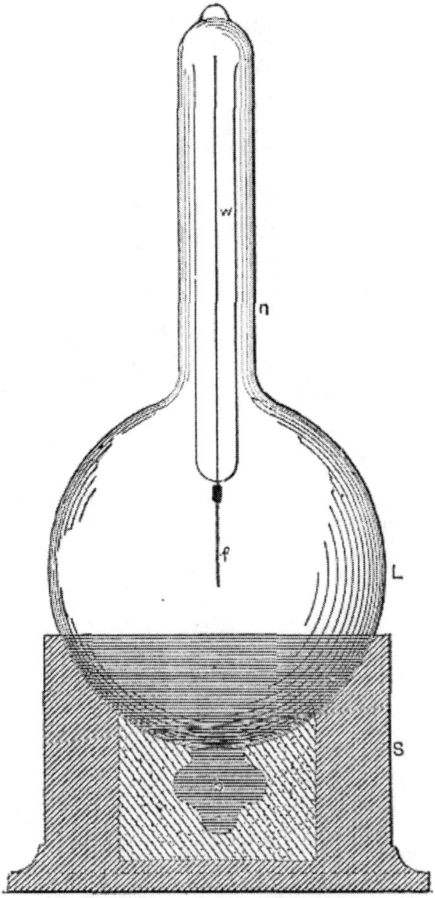

FIG. 24.—BULB WITHOUT LEADING-IN WIRE, SHOWING EFFECT OF PROJECTED MATTER.

In Fig. 24, for example, an experimental bulb L is shown, which is provided with a neck n on the top for the application of an external tinfoil coating, which may be connected to a body of larger surface.

Such a lamp as illustrated in Fig. 25 may also be lighted by connecting the tinfoil coating on the neck n to the terminal, and the leading-in wire w to an insulated plate. If the bulb stands in a socket upright, as shown in the cut, a shade of conducting material may be slipped in the neck n, and the action thus magnified.

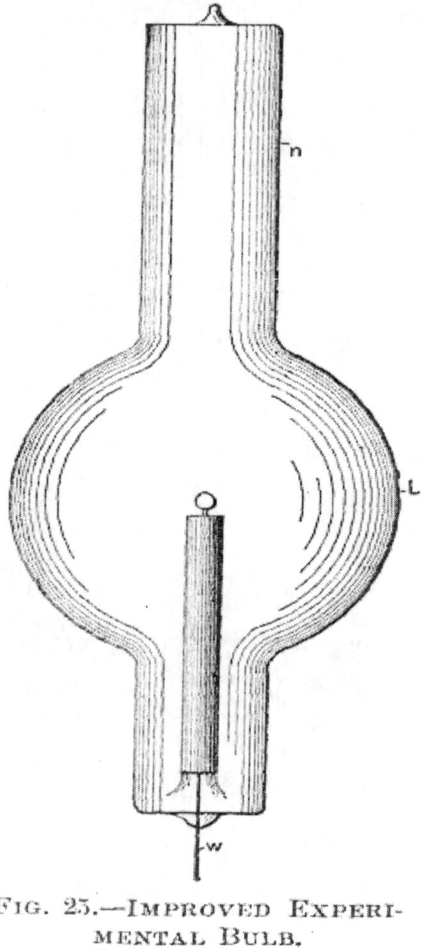


FIG. 25.—IMPROVED EXPERIMENTAL BULB.

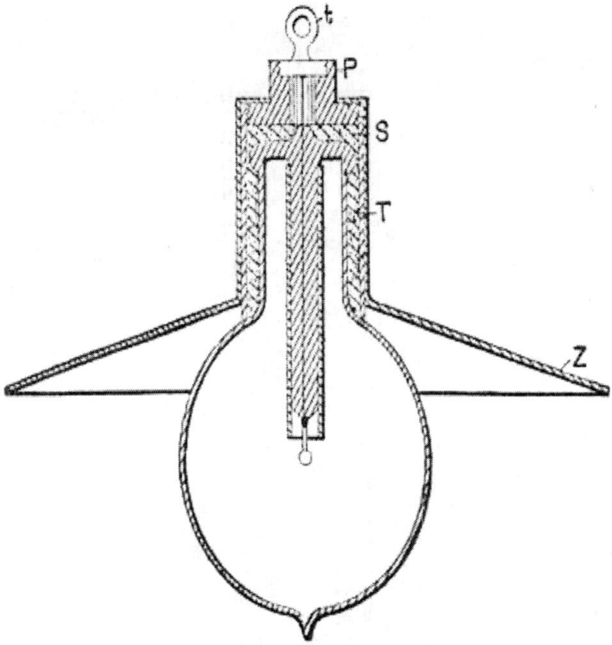


FIG. 26.—IMPROVED BULB WITH INTENSIFYING REFLECTOR.

A more perfected arrangement used in some of these bulbs is illustrated in Fig. 26. In this case the construction of the bulb is as shown and described before, when reference was made to Fig. 19. A zinc sheet Z, with a tubular extension T, is slipped over the metallic socket S. The bulb hangs downward from the terminal t, the zinc sheet Z, performing the double office of intensifier and reflector. The reflector is separated from the terminal t by an extension of the insulating plug P.

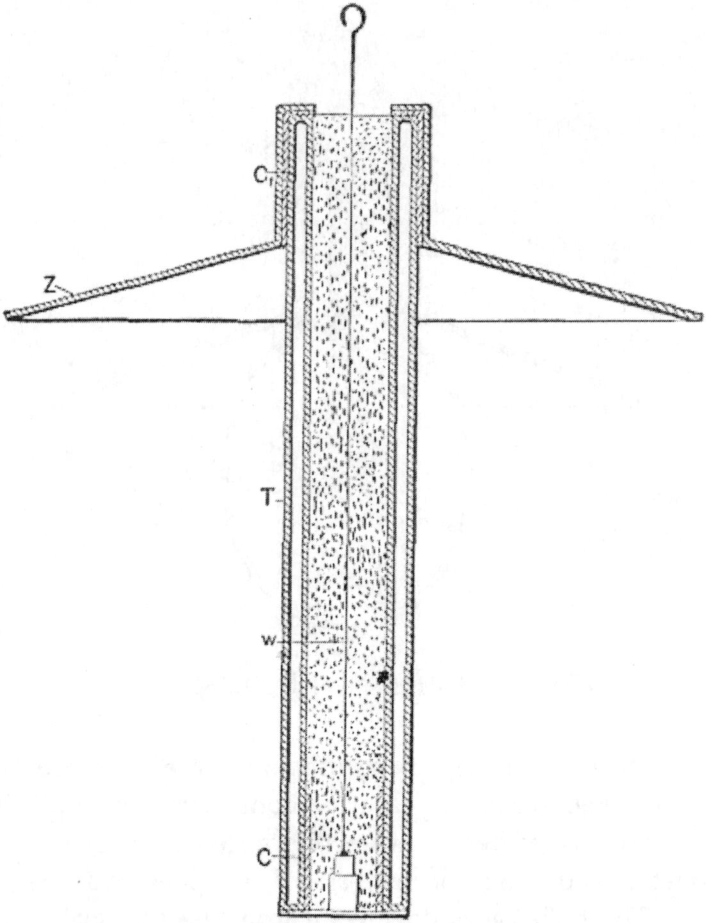


FIG. 27.—PHOSPHORESCENT TUBE WITH INTENSIFYING REFLECTOR.

A similar disposition with a phosphorescent tube is illustrated in Fig. 27. The tube T is prepared from two short tubes of a different diameter, which are sealed on the ends. On the lower end is placed an outside conducting coating C, which connects to the wire w. The wire has a hook on the upper end for suspension, and passes through the center of the inside tube, which is filled with some good and tightly packed insulator. On the outside of the upper end of the tube T is another conducting coating C1 upon which is slipped a metallic reflector Z, which should be separated by a thick insulation from the end of wire w.

The economical use of such a reflector or intensifier would require that all energy supplied to an air condenser should be recoverable, or, in other words, that there should not be any losses, neither in the gaseous medium nor through its action elsewhere. This is far from being so, but, fortunately, the losses may be reduced to anything desired. A few remarks are necessary on this subject, in order to make the experiences gathered in the course of these investigations perfectly clear.

Suppose a small helix with many well insulated turns, as in experiment Fig. 17, has one of its ends connected to one of the terminals of the induction coil, and the other to a metal plate, or, for the sake of simplicity, a sphere, insulated in space. When the coil is set to work, the potential of the sphere is alternated, and the small helix now behaves as though its free end were connected to the other terminal of the induction coil. If an iron rod be held within the small helix, it is quickly brought to a high temperature, indicating the passage of a strong current through the helix. How does the insulated sphere act in this case? It can be a condenser, storing and returning the energy supplied to it, or it can be a mere sink of energy, and the conditions of the experiment determine whether it is more one or the other. The sphere being charged to a high potential, it acts inductively upon the surrounding air, or whatever gaseous medium there might be. The molecules, or atoms, which are near the sphere are of course more attracted, and move through a greater distance than the farther ones. When the nearest molecules strike the sphere, they are repelled, and collisions occur at all distances within the inductive action of the sphere. It is now clear that, if the potential be steady, but little loss of energy can be caused in this way, for the molecules which are nearest to the sphere, having had an additional charge imparted to them by contact, are not attracted until they have parted, if not with all, at least with most of the additional charge, which can be accomplished only after a great many collisions. From the fact that with a steady potential there is but little loss in dry air, one must come to such a conclusion. When the potential of the sphere, instead of being steady, is alternating, the conditions are entirely different. In this case a rhythmical bombardment occurs, no matter whether the molecules after coming in contact with the sphere lose the imparted charge or not; what is more, if the charge is not lost, the impacts are only the more violent. Still if the frequency of the impulses be very small, the loss caused by the impacts and collisions would not be serious unless the potential were excessive. But when extremely high frequencies and more or less high potentials are used, the loss may be very great. The total energy lost per unit of time is proportionate to the product of the number of impacts per second, or the frequency and the energy lost in each impact. But the energy of an impact must be proportionate to the square of the electric density of the sphere, since the charge imparted to the molecule is proportionate to that density. I conclude from this that the total energy lost must be proportionate to the product of the frequency and the square of the electric density; but this law needs experimental confirmation. Assuming the preceding considerations to be true, then, by rapidly alternating the potential of a body immersed in an insulating gaseous medium, any amount of energy may be dissipated into space. Most of that energy then, I believe, is not dissipated in the form of long ether waves, propagated to considerable distance, as is thought most generally, but is consumed in the case of an insulated sphere, for example in impact and collisional losses that is, heat vibrations on the surface and in the vicinity of the sphere. To reduce the dissipation, it is necessary to work with a small electric density the smaller the higher the frequency.

But since, on the assumption before made, the loss is diminished with the square of the density, and since currents of very high frequencies involve considerable waste when transmitted through conductors, it follows that, on the whole, it is better to employ one wire than two. Therefore, if motors, lamps, or devices of any kind are perfected, capable of being advantageously operated by currents of extremely high frequency, economical reasons will make it advisable to use only one wire, especially if the distances are great.

When energy is absorbed in a condenser the same behaves as though its capacity were increased. Absorption always exists more or less, but generally it is small and of no consequence as long as the frequencies are not very great. In using extremely high frequencies, and, necessarily in such case, also high potentials, the absorption or, what is here meant more particularly by this term, the loss of energy due to the presence of a gaseous medium is an important factor to be considered, as the energy absorbed in the air condenser may be any fraction of the supplied energy. This would seem to make it very difficult to tell from the measured or computed capacity of an air condenser its actual capacity or vibration period, especially if the condenser is of very small surface and is charged to a very high potential. As many important results are dependent upon the correctness of the estimation of the vibration period, this subject demands the most careful scrutiny of other investigators. To reduce the probable error as much as possible in experiments of the kind alluded to, it is advisable to use spheres or plates of large surface, so as to make the density exceedingly small. Otherwise, when it is practicable, an oil condenser should be used in preference. In oil or other liquid dielectrics there are seemingly no such losses as in gaseous media. It being impossible to exclude entirely the gas in condensers with solid dielectrics, such condensers should be immersed in oil, for economical reasons if nothing else; they can then be strained to the utmost and will remain cool. In Leyden jars the loss due to air is comparatively small, as the tinfoil coatings are large, close together, and the charged surfaces not directly exposed; but when the potentials are very high, the loss may be more or less considerable at, or near, the upper edge of the foil, where the air is principally acted upon. If the jar be immersed in boiled-out oil, it will be capable of performing four times the amount of work which it can for any length of time when used in the ordinary way, and the loss will be inappreciable.

It should not be thought that the loss in heat in an air condenser is necessarily associated with the formation of *visible* streams or brushes. If a small electrode, enclosed in an unexhausted bulb, is connected to one of the terminals of the coil, streams can be seen to issue from the electrode and the air in the bulb is heated; if, instead of a small electrode, a large sphere is enclosed in the bulb, no streams are observed, still the air is heated.

Nor should it be thought that the temperature of an air condenser would give even an approximate idea of the loss in heat incurred, as in such case heat must be given off much more quickly, since there is, in addition to the ordinary radiation, a very active carrying away of heat by independent carriers going on, and since not only the apparatus, but the air at some distance from it is heated in consequence of the collisions which must occur.

Owing to this, in experiments with such a coil, a rise of temperature can be distinctly observed only when the body connected to the coil is very small. But with apparatus on a larger scale, even a body of considerable bulk would be heated, as, for instance, the body of a person; and I think that skilled physicians might make observations of utility in such experiments, which, if the apparatus were judiciously designed, would not present the slightest danger.

A question of some interest, principally to meteorologists, presents itself here. How does the earth behave? The earth is an air condenser, but is it a perfect or a very imperfect one a mere sink of energy? There can be little doubt that to such small disturbance as might be caused in an experiment the earth behaves as an almost perfect condenser. But it might be different when its charge is set in vibration by some sudden disturbance occurring in the heavens. In such case, as before stated, probably only little of the energy of the vibrations set up would be lost into space in the form of long ether radiations, but most of the energy, I think, would spend itself in molecular impacts and collisions, and pass off into space in the form of short heat, and possibly light, waves. As both the frequency of the vibrations of the charge and the potential are in all probability excessive, the energy converted into heat may be considerable. Since the density must be unevenly distributed, either in consequence of the irregularity of the earth's surface, or on account of the condition of the atmosphere in various places, the effect produced would accordingly vary from place to place. Considerable variations in the temperature and pressure of the atmosphere may in this manner be caused at any point of the surface of the earth. The variations may be gradual or very sudden, according to the nature of the general disturbance, and may produce rain and storms, or locally modify the weather in any way.

From the remarks before made one may see what an important factor of loss the air in the neighborhood of a charged surface becomes when the electric density is great and the frequency of the impulses excessive. But the action as explained implies that the air is insulating that is, that it is composed of independent carriers immersed in an insulating medium. This is the case only when the air is at something like ordinary or greater, or at extremely small, pressure. When the air is slightly rarefied and conducting, then true conduction losses occur also. In such case, of course, considerable energy may be dissipated into space even with a steady potential, or with impulses of low frequency, if the density is very great.

When the gas is at very low pressure, an electrode is heated more because higher speeds can be reached. If the gas around the electrode is strongly compressed, the displacements, and consequently the speeds, are very small, and the heating is insignificant. But if in such case the frequency could be sufficiently increased, the electrode would be brought to a high temperature as well as if the gas were at very low pressure; in fact, exhausting the bulb is only necessary because we cannot produce (and possibly not convey) currents of the required frequency.

Returning to the subject of electrode lamps, it is obviously of advantage in such a lamp to confine as much as possible the heat to the electrode by preventing the circulation of the gas in the bulb. If a very small bulb be taken, it would confine the heat better than a large one, but it might not be of sufficient capacity to be operated from the coil, or, if so, the glass might get too hot. A simple way to improve in this direction is to employ a globe of the required size, but to place a small bulb, the diameter of which is properly estimated, over the refractory button contained in the globe. This arrangement is illustrated in Fig. 28.

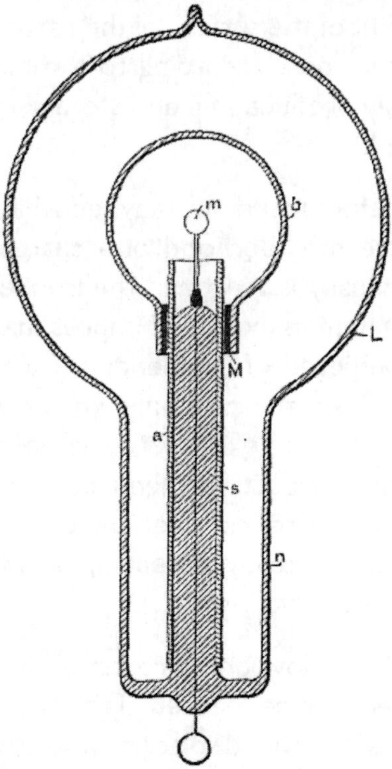


FIG. 28.—LAMP WITH AUXILIARY BULB FOR CONFINING THE ACTION TO THE CENTRE.

The globe L has in this case a large neck n, allowing the small bulb b to slip through. Otherwise, the construction is the same as shown in Fig. 18, for example. The small bulb is conveniently supported upon the stem s, carrying the refractory button m. It is separated from the aluminums tube a by several layers of mica M, in order to prevent the cracking of the neck by the rapid heating of the aluminum tube upon a sudden turning on of the current. The inside bulb should be as small as possible when it is desired to obtain light only by incandescence of the electrode. If it is desired to produce phosphorescence, the bulb should be larger, else it would be apt to get too hot, and the phosphorescence would cease. In this arrangement usually only, the small bulb shows phosphorescence, as there is practically no bombardment against the outer globe. In some of these bulbs constructed as illustrated in Fig. 28 the small tube was coated with phosphorescent paint, and beautiful effects were obtained. Instead of making the inside bulb large, in order to avoid undue heating, it answers the purpose to make the electrode m larger. In this case the bombardment is weakened by reason of the smaller electric density.

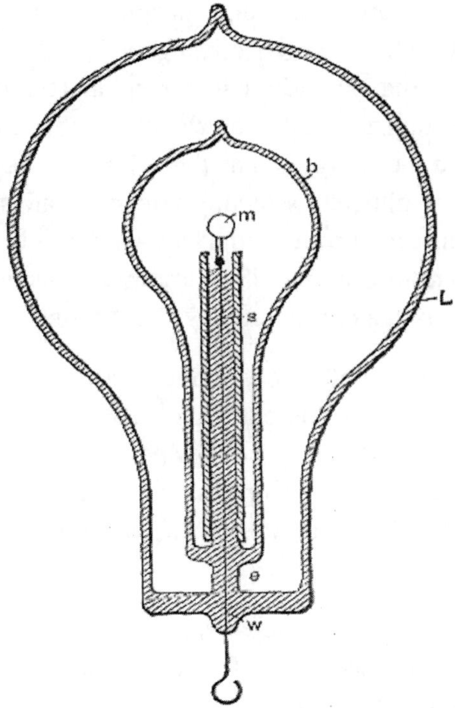


FIG. 29.—LAMP WITH INDEPENDENT AUXILIARY BULB.

Many bulbs were constructed on the plan illustrated in Fig. 29. Here a small bulb b, containing the refractory button m, upon being exhausted to a very high degree was sealed in a large globe L, which was then moderately exhausted and sealed off. The principal advantage of this construction was that it allowed of reaching extremely high vacua, and, at the same time use a large bulb. It was found, in the course of experiences with bulbs such as illustrated in Fig. 29, that it was well to make the stem s near the seal at every thick, and the leading-in wire w thin, as it occurred sometimes that the stem ate was heated and the bulb was cracked. Often the outer globe L was exhausted only just enough to allow the discharge to pass through, and the space between the bulbs appeared crimson, producing a curious effect. In some cases, when the exhaustion in globe L was very low, and the air good conducting, it was found necessary, in order to bring the button m to high incandescence, to place, preferably on the upper part of the neck of the globe, a tinfoil coating which was connected to an insulated body, to the ground, or to the other terminal of the coil, as the highly conducting air weakened the effect somewhat, probably by being acted upon inductively from the wire w, where it entered the bulb at e. Another difficulty which, however, is always present when the refractory button is mounted in a very small bulb existed in the construction illustrated in Fig. 29, namely, the vacuum in the bulb b would be impaired in a comparatively short time.

The chief idea in the two last described constructions was to confine the heat to the central portion of the globe by preventing the exchange of air. An advantage is secured but owing to the heating of the inside bulb and slow evaporation of the glass the vacuum is hard to maintain, even if the construction illustrated in Fig. 28 be chosen, in which both bulbs communicate.

But by far the better way, the ideal way would be to reach sufficiently high frequencies. The higher the frequency the slower would be the exchange of the air, and I think that a frequency may be reached at which there would be no exchange whatever of the air molecules around the terminal. We would then produce a flame in which there would be no carrying away of material, and a queer flame it would be, for it would be rigid! With such high frequencies the inertia of the particles would come into play. As the brush, or flame, would gain rigidity in virtue of the inertia of the particles, the exchange of the latter would be prevented. This would necessarily occur, for, the number of the impulses being augmented, the potential energy of each would diminish, so that finally only atomic vibrations could be set up, and the motion of translation through measurable space would cease. Thus, an ordinary gas burner connected to a source of rapidly alternating potential might have its efficiency augmented to a certain limit, and this for two reasons because of the additional vibration imparted, and because of a slowing down of the process of carrying off. But the renewal being rendered difficult, and renewal being necessary to maintain the *burner*, a continued increase of the frequency of the impulses, assuming they could be transmitted to and impressed upon the flame, would result in the "extinction" of the latter, meaning by this term only the cessation of the chemical process.

I think, however, that in the case of an electrode immersed in a fluid insulating medium, and surrounded by independent carriers of electric charges, which can be acted upon inductively, a sufficiently high frequency of the impulses would probably result in a gravitation of the gas all around toward the electrode. For this it would be only necessary to assume that the independent bodies are irregularly shaped; they would then turn toward the electrode their side of the greatest electric density, and this would be a position in which the fluid resistance to approach would be smaller than that offered to the receding.

The general opinion, I do not doubt, is that it is out of the question to reach any such frequencies as might assuming some of the views before expressed to be true, produce any of the results which I have pointed out as mere possibilities. This may be so, but in the course of these investigations, from the observation of many phenomena I have gained the conviction that these frequencies would be much lower than one is apt to estimate at first. In a flame we set up light vibrations by causing molecules, or atoms, to collide. But what is the ratio of the frequency of the collisions and that of the vibrations set up? Certainly, it must be incomparably smaller than that of the knocks of the bell and the sound vibrations, or that of the discharges and the oscillations of the condenser. We may cause the molecules of the gas to collide by the use of alternate electric impulses of high frequency, and so we may imitate the process in a flame; and from experiments with frequencies which we are now able to obtain, I think that the result is producible with impulses which are transmissible through a conductor.

In connection with thoughts of a similar nature, it appeared to me of great interest to demonstrate the rigidity of a vibrating gaseous column. Although with such low frequencies as, say 10,000 per second, which I was able to obtain without difficulty from a specially constructed alternator, the task looked discouraging at first, I made a series of experiments. The trials with air at ordinary pressure led to no result, but with air moderately rarefied I obtain what I think to be an unmistakable experimental evidence of the property sought for. As a result of this kind might lead able investigators to conclusions of importance, I will describe one of the experiments performed.

It is well known that when a tube is slightly exhausted the discharge may be passed through it in the form of a thin luminous thread. When produced with currents of low frequency, obtained from a coil operated as usual, this thread is inert. If a magnet be approached to it, the part near the same is attracted or repelled, according to the direction of the lines of force of the magnet. It occurred to me that if such a thread would be produced with currents of very high frequency, it should be more or less rigid, and as it was visible it could be easily studied. Accordingly, I prepared a tube about 1 inch in diameter and 1 meter long, with outside coating at each end. The tube was exhausted to a point at which by a little working the thread discharge could be obtained. It must be remarked here that the general aspect of the tube, and the degree of exhaustion, are quite different than when ordinary low frequency currents are used. As it was found preferable to work with one terminal, the tube prepared was suspended from the end of a wire connected to the terminal, the tinfoil coating being connected to the wire, and to the lower coating sometimes a small, insulated plate was attached. When the thread was formed it extended through the upper part of the tube and lost itself in the lower end. If it possessed rigidity it resembled, not exactly an elastic cord stretched tight between two supports, but a cord suspended from a height with a small weight attached at the end. When the finger or a magnet was approached to the upper end of the luminous thread, it could be brought locally out of position by electrostatic or magnetic action; and when the disturbing object was very quickly removed, an analogous result was produced, as though a suspended cord would be displaced and quickly released near the point of suspension. In doing this the luminous thread was set in vibration, and two very sharply marked nodes, and a third indistinct one, were formed. The vibration, once set up, continued for fully eight minutes, dying gradually out. The speed of the vibration often varied perceptibly, and it could be observed that the electrostatic attraction of the glass affected the vibrating thread; but it was clear that the electrostatic action was not the cause of the vibration, for the thread was most generally stationary, and could always be set in vibration by passing the finger quickly near the upper part of the tube. With a magnet the thread could be split in two and both parts vibrated. By approaching the hand to the lower coating of the tube, or insulated plate if attached, the vibration was quickened; also, as far as I could see, by raising the potential or frequency. Thus, either increasing the frequency or passing a stronger discharge of the same frequency corresponded to a tightening of the cord. I did not obtain any experimental evidence with condenser discharges. A luminous band excited in a bulb by repeated discharges of a Leyden jar must possess rigidity, and if deformed and suddenly released should vibrate. But probably the amount of vibrating matter is so small that in spite of the extreme speed the inertia cannot prominently assert itself. Besides, the observation in such a case is rendered extremely difficult on account of the fundamental vibration.

The demonstration of the fact which still needs better experimental confirmation that a vibrating gaseous column possesses rigidity, might greatly modify the views of thinkers. When with low frequencies and insignificant potentials indications of that property may be noted, how must a gaseous medium behave under the influence of enormous electrostatic stresses which may be active in the interstellar space, and which may alternate with inconceivable rapidity? The existence of such an electrostatic, rhythmically throbbing force of a vibrating electrostatic field would show a possible way how solids might have formed from the ultra-gaseous uterus, and how transverse and all kinds of vibrations may be transmitted through a gaseous medium filling all space. Then, ether might be a true fluid, devoid of rigidity, and at rest, it being merely necessary as a connecting link to enable interaction. What determines the rigidity of a body? It must be the speed and the amount of moving matter. In a gas the speed may be considerable, but the density is exceedingly small; in a liquid the speed would be likely to be small, though the density may be considerable; and in both cases the inertia resistance offered to displacement is practically *nil*. But place a gaseous (or liquid) column in an intense, rapidly alternating electrostatic field, set the particles vibrating with enormous speeds, then the inertia resistance asserts itself. A body might move with more or less freedom through the vibrating mass, but as a whole it would be rigid.

There is a subject which I must mention in connection with these experiments: it is that of high vacua. This is a subject the study of which is not only interesting, but useful, for it may lead to results of great practical importance. In commercial apparatus, such as incandescent lamps, operated from ordinary systems of distribution, a much higher vacuum than obtained at present would not secure a very great advantage. In such a case the work is performed on the filament and the gas is little concerned; the improvement, therefore, would be but trifling. But when we begin to use very high frequencies and potentials, the action of the gas becomes all important, and the degree of exhaustion materially modifies the results. As long as ordinary coils, even very large ones, were used, the study of the subject was limited, because just at a point when it became most interesting it had to be interrupted on account of the "non-striking" vacuum being reached. But presently we are able to obtain from a small disruptive discharge coil potentials much higher than even the largest coil was capable of giving, and what is more, we can make the potential alternate with great rapidity. Both of these results enable us now to pass a luminous discharge through almost any vacua obtainable, and the field of our investigations is greatly extended. Think we as we may, of all the possible directions to develop a practical illuminant, the line of high vacua seems to be the most promising at present. But to reach extreme vacua the appliances must be much more improved, and ultimate perfection will not be attained until we shall have discarded the mechanical and perfected an *electrical* vacuum pump. Molecules and atoms can be thrown out of a bulb under the action of an enormous potential: *this* will be the principle of the vacuum pump of the future. For the present, we must secure the best results we can with mechanical appliances. In this respect, it might not be out of the way to say a few words about the method of, and apparatus for, producing excessively high degrees of exhaustion of which I have availed myself in the course of these investigations. It is very probable that other experimenters have used similar arrangements; but as it is possible that there may be an item of interest in their description, a few remarks, which will render this investigation more complete, might be permitted.

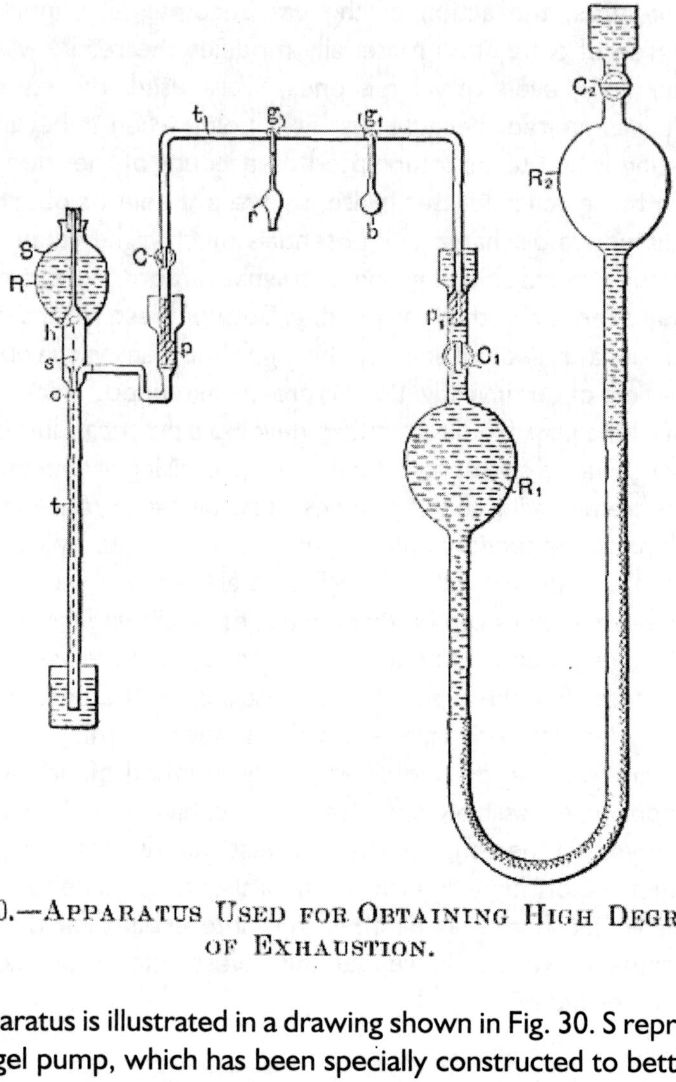


FIG. 30.—APPARATUS USED FOR OBTAINING HIGH DEGREES OF EXHAUSTION.

The apparatus is illustrated in a drawing shown in Fig. 30. S represents a Sprengel pump, which has been specially constructed to better suit the work required. The stopcock, which is usually employed has been omitted, and instead of it a hollow stopper s has been fitted in the neck of the reservoir R. This stopper has a small hole h, through which the mercury descends; the size of the outlet o being properly determined with respect to the section of the fall tube t, which is sealed to the reservoir instead of being connected to it in the usual manner. This arrangement overcomes the imperfections and troubles which often arise from the use of the stopcock on the reservoir and the connection of the latter with the fall tube.

The pump is connected through a U-shaped tube t to a very large reservoir R1. Especial care was taken in fitting the grinding surfaces of the stoppers p and p1, and both of these and the mercury caps above them were made exceptionally long. After the U-shaped tube was fitted and put in place, it was heated, so as to soften and take off the strain resulting from imperfect fitting. The U-shaped tube was provided with a stopcock C, and two ground connections g and g1 one for a small bulb b, usually containing caustic potash, and the other for the receiver r, to be exhausted.

The reservoir R1 was connected by means of a rubber tube to a slightly larger reservoir R2, each of the two reservoirs being provided with a stopcock C1 and C2, respectively. The reservoir R1 could be raised and lowered by a wheel and rack, and the range of its motion was so determined that when it was filled with mercury and the stopcock C2 closed, so as to form a Torricellian vacuum in it when raised, it could be lifted so high that the mercury in reservoir R1 would stand a little above stopcock C1; and when this stopcock was closed and the reservoir R2 descended, so as to form a Torricellian vacuum in reservoir R1, it could be lowered so far as to completely empty the latter, the mercury filling the reservoir R2 up to a little above stopcock C2.

The capacity of the pump and of the connections was taken as small as possible relatively to the volume of reservoir R1, since, of course, the degree of exhaustion depended upon the ratio of these quantities.

With this apparatus I combined the usual means indicated by former experiments for the production of very high vacua. In most of the experiments it was convenient to use caustic potash. I may venture to say, in regard to its use, that much time is saved, and a more perfect action of the pump insured by fusing and boiling the potash as soon as, or even before, the pump settles down. If this course is not followed the sticks, as ordinarily employed, may give moisture off at a certain very slow rate, and the pump may work for many hours without reaching a very high vacuum. The potash was heated either by a spirit lamp or by passing a discharge through it, or by passing a current through a wire contained in it. The advantage in the latter case was that the heating could be more rapidly repeated.

Generally, the process of exhaustion was the following: At the start, the stop-cocks C and C1 being open, and all other connections closed, the reservoir R2 was raised so far that the mercury filled the reservoir R1 and a part of the narrow connecting U-shaped tube. When the pump was set to work, the mercury would, of course, quickly rise in the tube, and reservoir R2 was lowered, the experimenter keeping the mercury at about the same level. The reservoir R2 was balanced by a long spring which facilitated the operation, and the friction of the parts was generally sufficient to keep it almost in any position. When the Sprengel pump had done its work, the reservoir R2 was further lowered and the mercury descended in R1 and filled R2, whereupon stopcock C2 was closed. The air adhering to the walls of R1 and that absorbed by the mercury was carried off, and to free the mercury of all air the reservoir R2 was for a long time worked up and down. During this process some air, which would gather below stopcock C2, was expelled from R2 by lowering it far enough and opening the stopcock, closing the latter again before raising the reservoir. When all the air had been expelled from the mercury, and no air would gather in R2 when it was lowered, the caustic potash was resorted to. The reservoir R2 was now again raised until the mercury in R1 stood above stopcock C1. The caustic potash was fused and boiled, and the moisture partly carried off by the pump and partly re-absorbed; and this process of heating and cooling was repeated many times, and each time, upon the moisture being absorbed or carried off, the reservoir R2 was for a long time raised and lowered. In this manner all the moisture was carried off from the mercury, and both the reservoirs were in proper condition to be used. The reservoir R2 was then again raised to the top, and the pump was kept working for a long time. When the highest vacuum obtainable with the pump had been reached the potash bulb was usually wrapped with cotton which was sprinkled with ether so as to keep the potash at a very low temperature, then the reservoir R2 was lowered, and upon reservoir R1 being emptied the receiver r was quickly sealed up.

When a new bulb was put on, the mercury was always raised above stopcock C1 which was closed, so as to always keep the mercury and both the reservoirs in fine condition, and the mercury was never withdrawn from R1 except when the pump had reached the highest degree of exhaustion. It is necessary to observe this rule if it is desired to use the apparatus to advantage.

By means of this arrangement I was able to proceed very quickly, and when the apparatus was in perfect order it was possible to reach the phosphorescent stage in a small bulb in less than 15 minutes, which is certainly very quick work for a small laboratory arrangement requiring all in all about 100 pounds of mercury. With ordinary small bulbs the ratio of the capacity of the pump, receiver, and connections, and that of reservoir R was about 1-20, and the degrees of exhaustion reached were necessarily very high, though I am unable to make a precise and reliable statement how far the exhaustion was carried.

What impresses the investigator most in the course of these experiences is the behavior of gases when subjected to great rapidly alternating electrostatic stresses. But he must remain in doubt as to whether the effects observed are due wholly to the molecules, or atoms, of the gas which chemical analysis discloses to us, or whether there enters into play another medium of a gaseous nature, comprising atoms, or molecules, immersed in a fluid pervading the space. Such a medium surely must exist, and I am convinced that, for instance, even if air were absent, the surface and neighborhood of a body in space would be heated by rapidly alternating the potential of the body; but no such heating of the surface or neighborhood could occur if all free atoms were removed and only a homogeneous, incompressible, and elastic fluid such as ether is supposed to be would remain, for then there would be no impacts, no collisions. In such a case, as far as the body itself is concerned, only frictional losses in the inside could occur.

It is a striking fact that the discharge through a gas is established with ever increasing freedom as the frequency of the impulses is augmented. It behaves in this respect quite contrarily to a metallic conductor. In the latter the impedance enters prominently into play as the frequency is increased, but the gas acts much as a series of condensers would: the facility with which the discharge passes through seems to depend on the rate of change of potential. If it acts so, then in a vacuum tube even of great length, and no matter how strong the current, self-induction could not assert itself to any appreciable degree. We have, then, as far as we can now see, in the gas a conductor which is capable of transmitting electric impulses of any frequency which we may be able to produce. Could the frequency be brought high enough, then a queer system of electric distribution, which would be likely to interest gas companies, might be realized: metal pipes filled with gas the metal being the insulator, the gas the conductor supplying phosphorescent bulbs, or perhaps devices as yet un-invented. It is certainly possible to take a hollow core of copper, rarefy the gas in the same, and bypassing impulses of sufficiently high frequency through a circuit around it, bring the gas inside to a high degree of incandescence; but as to the nature of the forces there would be considerable uncertainty, for it would be doubtful whether with such impulses the copper core would act as a static screen. Such paradoxes and apparent impossibilities we encounter at every step in this line of work, and therein lies, to a great extent, the claim of the study.

I have here a short and wide tube which is exhausted to a high degree and covered with a substantial coating of bronze, the coating allowing barely the light to shine through. A metallic clasp, with a hook for suspending the tube, is fastened around the middle portion of the latter, the clasp being in contact with the bronze coating. I now want to light the gas inside by suspending the tube on a wire connected to the coil. Anyone who would try the experiment for the first time, not having any previous experience, would probably take care to be quite alone when making the trial, for fear that he might become the joke of his assistants. Still, the bulb lights in spite of the metal coating, and the light can be distinctly perceived through the latter. A long tube covered with aluminum bronze lights when held in one hand, the other touching the terminal of the coil quite powerfully. It might be objected that the coatings are not sufficiently conducting; still, even if they were highly resistant, they ought to screen the gas. They certainly screen it perfectly in a condition of rest, but not by far perfectly when the charge is surging in the coating. But the loss of energy which occurs within the tube, notwithstanding the screen, is occasioned principally by the presence of the gas. Were we to take a large hollow metallic sphere and fill it with a perfect incompressible fluid dielectric, there would be no loss inside of the sphere, and consequently the inside might be considered as perfectly screened, though the potential be very rapidly alternating? Even were the sphere filled with oil, the loss would be incomparably smaller than when the fluid is replaced by a gas, for in the latter case the force produces displacements; that means impact and collisions in the inside.

No matter what the pressure of the gas may be, it becomes an important factor in the heating of a conductor when the electric density is great and the frequency very high. That in the heating of conductors by lightning discharges air is an element of great importance, is almost as certain as an experimental fact. I may illustrate the action of the air by the following experiment: I take a short tube which is exhausted to a moderate degree and has a platinum wire running through the middle from one end to the other. I pass a steady or low frequency current through the wire, and it is heated uniformly in all parts. The heating here is due to conduction, or frictional losses, and the gas around the wire has as far as we can see no function to perform. But now let me pass sudden discharges, or a high frequency current, through the wire. Again, the wire is heated, this time principally on the ends and least in the middle portion; and if the frequency of the impulses, or the rate of change, is high enough, the wire might as well be cut in the middle as not, for practically all the heating is due to the rarefied gas. Here the gas might only act as a conductor of no impedance diverting the current from the wire as the impedance of the latter is enormously increased, and merely heating the ends of the wire by reason of their resistance to the passage of the discharge. But it is not at all necessary that the gas in the tube should he conducting; it might be at an extremely low pressure, still the ends of the wire would be heated as, however, is ascertained by experience only the two ends would in such, case not be electrically connected through the gaseous medium. Now what with these frequencies and potentials occurs in an exhausted tube occurs in the lightning discharges at ordinary pressure. We only need remember one of the facts arrived at in the course of these investigations, namely, that to impulses of very high frequency the gas at ordinary pressure behaves much in the same manner as though it were at moderately low pressure. I think that in lightning discharges frequently wires or conducting objects are volatilized merely because air is present and that, were the conductor immersed in an insulating liquid, it would be safe, for then the energy would have to spend itself somewhere else. From the behavior of gases to sudden impulses of high potential I am led to conclude that there can be no surer way of diverting a lightning discharge than by affording it a passage through a volume of gas, if such a thing can be done in a practical manner.

There are two more features upon which I think it necessary to dwell in connection with these experiments the "radiant state" and the "non-striking vacuum."

Anyone who has studied Crookes' work must have received the impression that the "radiant state" is a property of the gas inseparably connected with an extremely high degree of exhaustion. But it should be remembered that the phenomena observed in an exhausted vessel are limited to the character and capacity of the apparatus which is made use of. I think that in a bulb a molecule, or atom, does not precisely move in a straight line because it meets no obstacle, but because the velocity imparted to it is sufficient to propel it in a sensibly straight line. The mean free path is one thing, but the velocity, the energy associated with the moving body is another, and under ordinary circumstances I believe that it is a mere question of potential or speed. A disruptive discharge coil, when the potential is pushed very far, excites phosphorescence and projects shadows, at comparatively low degrees of exhaustion. In a lightning discharge, matter moves in straight lines as ordinary pressure when the mean free path is exceedingly small, and frequently images of wires or other metallic objects have been produced by the particles thrown off in straight lines.

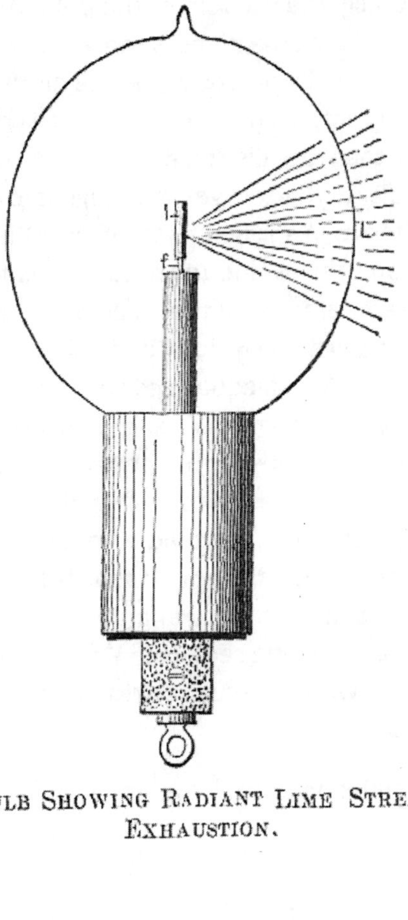

FIG. 31.—BULB SHOWING RADIANT LIME STREAM AT LOW EXHAUSTION.

I have prepared a bulb to illustrate by an experiment the correctness of these assertions. In a globe L (Fig. 31) I have mounted upon a lamp filament f a piece of lime l. The lamp filament is connected with a wire which leads into the bulb, and the general construction of the latter is as indicated in Fig. 19, before described. The bulb being suspended from a wire connected to the terminal of the coil, and the latter being set to work, the lime piece l and the projecting parts of the filament f are bombarded. The degree of exhaustion is just such that with the potential the coil is capable of giving phosphorescence of the glass is produced but disappears as soon as the vacuum is impaired. The lime containing moisture, and moisture being given off as soon as heating occurs, the phosphorescence lasts only for a few moments. When the lime has been sufficiently heated, enough moisture has been given off to impair materially the vacuum of the bulb. As the bombardment goes on, one point of the lime piece is more heated than other points, and the result is that finally practically all the discharge passes through that point which is intensely heated, and a white stream of lime particles (Fig. 31) then breaks forth from that point. This stream is composed of "radiant" matter, yet the degree of exhaustion is low. But the particles move in straight lines because the velocity imparted to them is great, and this is due to three causes to the great electric density, the high temperature of the small point, and the fact that the particles of the lime are easily torn and thrown off far more easily than those of carbon. With frequencies such as we are able to obtain, the particles are bodily thrown off and projected to a considerable distance; but with sufficiently high frequencies no such thing would occur in such case only a stress would spread, or a vibration would be propagated through the bulb. It would be out of the question to reach any such frequency on the assumption that the atoms move with the speed of light; but I believe that such a thing is impossible; for this an enormous potential would be required. With potentials which we are able to obtain, even with a disruptive discharge coil, the speed must be quite insignificant.

As to the "non-striking vacuum," the point to be noted is that it can occur only with low frequency impulses, and it is necessitated by the impossibility of carrying off enough energy with such impulses in high vacuum since the few atoms which are around the terminal upon coming in contact with the same are repelled and kept at a distance for a comparatively long period of time, and not enough work can be performed to render the effect perceptible to the eye. If the difference of potential between the terminals is raised, the dielectric breaks down. But with very high frequency impulses there is no necessity for such breaking down, since any amount of work can be performed by continually agitating the atoms in the exhausted vessel, provided the frequency is high enough. It is easy to reach even with frequencies obtained from an alternator as here used, a stage at which the discharge does not pass between two electrodes in a narrow tube, each of these being connected to one of the terminals of the coil, but it is difficult to reach a point at which a luminous discharge would not occur around each electrode.

A thought which naturally presents itself in connection with high frequency currents, is to make use of their powerful electro-dynamic inductive action to produce light effects in a sealed glass globe. The leading-in wire is one of the defects of the present incandescent lamp, and if no other improvement were made, that imperfection at least should be done away with. Following this thought, I have carried on experiments in various directions, of which some were indicated in my former paper. I may here mention one or two more lines of experiment which have been followed up.

Many bulbs were constructed as shown in Fig. 32 and Fig. 33.

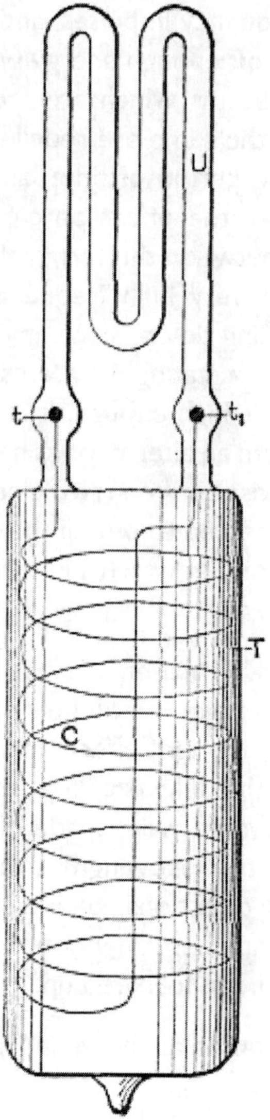


FIG. 32.—ELECTRO-DYNAMIC INDUCTION TUBE.

In Fig. 32 a wide tube T was sealed to a smaller W-shaped tube U, of phosphorescent glass. In the tube T was placed a coil C of aluminum wire, the ends of which were provided with small spheres t and t1 of aluminum and reached into the U tube. The tube T was slipped into a socket containing a primary coil through which usually the discharges of Leyden jars were directed, and the rarefied gas in the small U tube was excited to strong luminosity by the high-tension currents induced in the coil C. When Leyden jar discharges were used to induce currents in the coil C, it was found necessary to pack the tube T tightly with insulating powder, as a discharge would occur frequently between the turns of the coil, especially when the primary was thick and the air gap, through which the jars discharged, large, and no little trouble was experienced in this way.

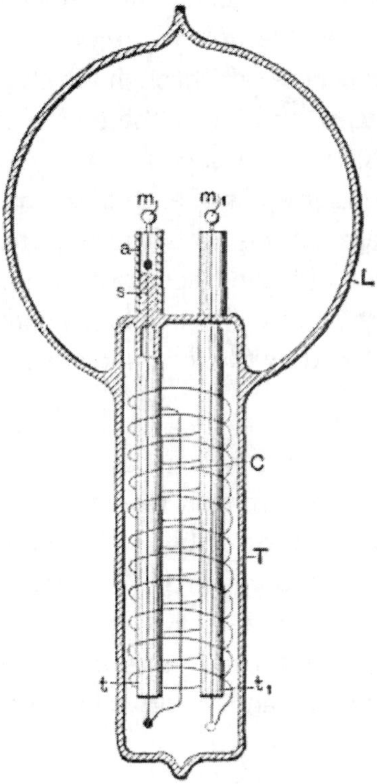


FIG. 33.—ELECTRO-DYNAMIC INDUCTION LAMP.

In Fig. 33 is illustrated another form of the bulb constructed. In this case a tube T is sealed to a globe L. The tube contains a coil C, the ends of which pass through two small glass tubes t and t1, which are sealed to the tube T. Two refractory buttons m and m1 are mounted on lamp filaments which are fastened to the ends of the wires passing through the glass tubes t and t1. Generally, in bulbs made on this plan the globe L communicated with the tube T. For this purpose, the ends of the small tubes t and t1 were just a trifle heated in the burner, merely to hold the wires, but not to interfere with the communication. The tube T, with the small tubes, wires through the same, and the refractory buttons m and m1, was first prepared, and then sealed to globe L, whereupon the coil C was slipped in, and the connections made to its ends. The tube was then packed with insulating powder, jamming the latter as tight as possible up to very nearly the end, then it was closed and only a small hole left through which the remainder of the powder was introduced, and finally the end of the tube was closed. Usually in bulbs constructed as shown in Fig. 33 an aluminum tube a was fastened to the upper end s of each of the tubes t and t1, in order to protect that end against the heat. The buttons m and m1 could be brought to any degree of incandescence by passing the discharges of Leyden jars around the coil C. In such bulbs with two buttons a very curious effect is produced by the formation of the shadows of each of the two buttons.

Another line of experiment, which has been assiduously followed, was to induce by electro-dynamic induction a current or luminous discharge in an exhausted tube or bulb. This matter has received such able treatment at the hands of Prof. J.J. Thomson that I could add but little to what he has made known, even had I made it the special subject of this lecture. Still, since experiences in this line have gradually led me to the present views and results, a few words must be devoted here to this subject.

It has occurred, no doubt, to many that as a vacuum tube is made longer the electromotive force per unit length of the tube, necessary to pass a luminous discharge through the latter, gets continually smaller; therefore, if the exhausted tube be made long enough, even with low frequencies a luminous discharge could be induced in such a tube closed upon itself. Such a tube might be placed around a ball or on a ceiling, and at once a simple appliance capable of giving considerable light would be obtained. But this would be an appliance hard to manufacture and extremely unmanageable. It would not do to make the tube up of small lengths, because there would be with ordinary frequencies considerable loss in the coatings, and besides, if coatings were used, it would be better to supply the current directly to the tube by connecting the coatings to a transformer. But even if all objections of such nature were removed, still, with low frequencies the light conversion itself would be inefficient, as I have before stated. In using extremely high frequencies the length of the secondary in other words, the size of the vessel can be reduced as far as desired, and the efficiency of the light conversion is increased, provided that means are invented for efficiently obtaining such high frequencies. Thus, one is led, from theoretical and practical considerations to the use of high frequencies, and this means high electromotive forces and small currents in the primary. When he works with condenser charges, and they are the only means up to the present known for reaching these extreme frequencies he gets to electromotive forces of several thousands of volts per turn of the primary. He cannot multiply the electro-dynamic inductive effect by taking more turns in the primary, for he arrives at the conclusion that the best way is to work with one single turn though he must sometimes depart from this rule, and he must get along with whatever inductive effect he can obtain with one turn. But before he has long experimented with the extreme frequencies required to set up in a small bulb an electromotive force of several thousands of volts he realizes the great importance of electrostatic effects, and these effects grow relatively to the electro-dynamic in significance as the frequency is increased.

Now, if anything is desirable in this case, it is to increase the frequency, and this would make it still worse for the electro-dynamic effects. On the other hand, it is easy to exalt the electrostatic action as far as one likes by taking more turns on the secondary or combining self-induction and capacity to raise the potential. It should also be remembered that, in reducing the current to the smallest value and increasing the potential, the electric impulses of high frequency can be more easily transmitted through a conductor.

These and similar thoughts determined me to devote more attention to the electrostatic phenomena, and to endeavor to produce potentials as high as possible and alternating as fast as they could be made to alternate. I then found that I could excite vacuum tubes at considerable distance from a conductor connected to a properly constructed coil, and that I could, by converting the oscillatory current of a condenser to a higher potential, establish electrostatic alternating fields which acted through the whole extent of a room, lighting up a tube no matter where it was held in space. I thought I recognized that I had made a step-in advance, and I have persevered in this line; but I wish to say that I share with all lovers of science and progress the one and only desire to reach a result of utility to men in any direction to which thought, or experiment may lead me. I think that this departure is the right one, for I cannot see, from the observation of the phenomena which manifest themselves as the frequency is increased, what there would remain to act between two circuits conveying, for instance, impulses of several hundred millions per second, except electrostatic forces. Even with such trifling frequencies the energy would be practically all potential, and my conviction has grown strong that, to whatever kind of motion light may be due, it is produced by tremendous electrostatic stresses vibrating with extreme rapidity.

Of all these phenomena observed with currents, or electric impulses, of high frequency, the most fascinating for an audience are certainly those which are noted in an electrostatic field acting through considerable distance, and the best an unskilled lecturer can do is to begin and finish with the exhibition of these singular effects. I take a tube in the hand and move it about, and it is lighted wherever I may hold it, throughout space the invisible forces act. But I may take another tube and it might not light, the vacuum being very high. I excite it by means of a disruptive discharge coil, and now it will light in the electrostatic field. I may put it away for a few weeks or months, still it retains the faculty of being excited. What change have I produced in the tube in the act of exciting it? If a motion imparted to the atoms, it is difficult to perceive how it can persist so long without being arrested by frictional losses; and if a strain exerted in the dielectric, such as a simple electrification would produce, it is easy to see how it may persist indefinitely, but very difficult to understand why such a condition should aid the excitation when we have to deal with potentials which are rapidly alternating.

Since I have exhibited these phenomena for the first time, I have obtained some other interesting effects. For instance, I have produced the incandescence of a button, filament, or wire enclosed in a tube. To get to this result it was necessary to economize the energy which is obtained from the field and direct most of it on the small body to be rendered incandescent. At the beginning the task appeared difficult, but the experiences gathered permitted me to reach the result easily. In Fig. 34 and Fig. 35 two such tubes are illustrated which are prepared for the occasion.

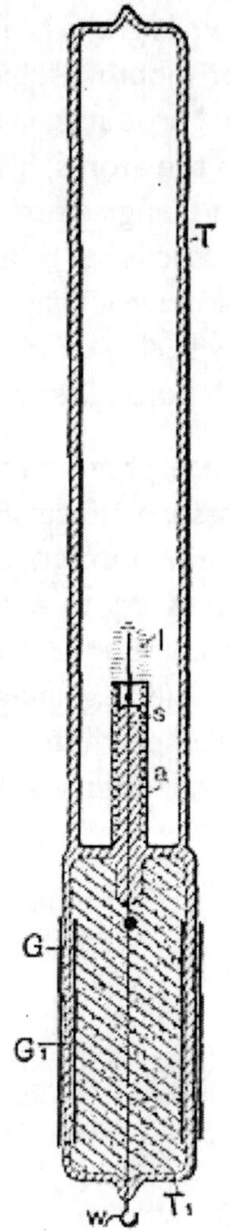


FIG. 34.—TUBE WITH FILAMENT RENDERED INCANDESCENT IN AN ELECTRO STATIC FIELD.

In Fig. 34 a short tube T1, sealed to another long tube T, is provided with a stem s, with a platinum wire sealed in the latter. A very thin lamp filament l is fastened to this wire, and connection to the outside is made through a thin copper wire w. The tube is provided with outside and inside coatings, C and C1 respectively, and is filled as far as the coatings reach with conducting, and the space above with insulating powder. These coatings are merely used to enable me to perform two experiments with the tube namely, to produce the effect desired either by direct connection of the body of the experimenter or of another body to the wire w, or by acting inductively through the glass. The stem s is provided with an aluminum tube a, for purposes before explained, and only a small part of the filament reaches out of this tube. By holding the tube T1 anywhere in the electrostatic field the filament is rendered incandescent.

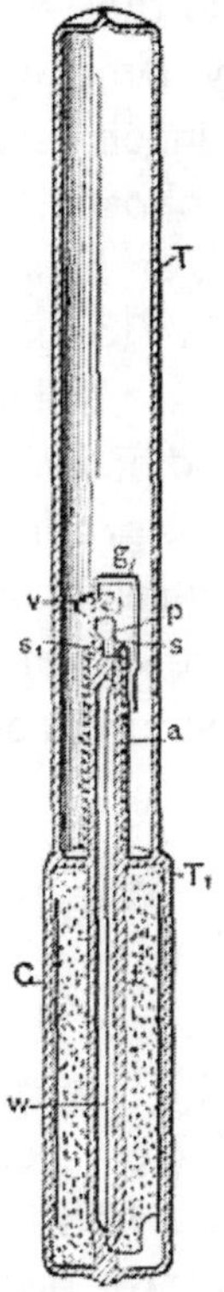


FIG. 35. — CROOKES' EXPERIMENT IN ELECTROSTATIC FIELD.

A more interesting piece of apparatus is illustrated in Fig. 35. The construction is the same as before, only instead of the lamp filament a small platinum wire p, sealed in a stem s, and bent above it in a circle, is connected to the copper wire w, which is joined to an inside coating C. A small stem s l is provided with a needle, on the point of which is arranged to rotate very freely a very light fan of mica v. To prevent the fan from falling out, a thin stem of glass g is bent properly and fastened to the aluminum tube. When the glass tube is held anywhere in the electrostatic field the platinum wire becomes incandescent, and the mica vanes are rotated very fast.

Intense phosphorescence may be excited in a bulb by merely connecting it to a plate within the field, and the plate need not be any larger than an ordinary lamp shade. The phosphorescence excited with these currents is incomparably more powerful than with ordinary apparatus. A small phosphorescent bulb, when attached to a wire connected to a coil, emits sufficient light to allow reading ordinary print at a distance of five to six paces. It was of interest to see how some of the phosphorescent bulbs of Professor Crookes would behave with these currents, and he has had the kindness to lend me a few for the occasion. The effects produced are magnificent, especially by the sulphide of calcium and sulphide of zinc. From the disruptive discharge coil, they glow intensely merely by holding them in the hand and connecting the body to the terminal of the coil.

To whatever results investigations of this kind may lead, their chief interest lies for the present in the possibilities they offer for the production of an efficient illuminating device. In no branch of electric industry is an advance more desired than in the manufacture of light. Every thinker, when considering the barbarous methods employed, the deplorable losses incurred in our best systems of light production, must have asked himself, what is likely to be the light of the future? Is it to be an incandescent solid, as in the present lamp, or an incandescent gas, or a phosphorescent body, or something like a burner, but incomparably more efficient?

There is little chance to perfect a gas burner; not, perhaps, because human ingenuity has been bent upon that problem for centuries without a radical departure having been made, though this argument is not devoid of force-but because in a burner the higher vibrations can never be reached except by passing through all the low ones. For how is a flame produced unless by a fall of lifted weights? Such process cannot be maintained without renewal, and renewal is repeated passing from low to high vibrations. One way only seems to be open to improve a burner, and that is by trying to reach higher degrees of incandescence. Higher incandescence is equivalent to a quicker vibration; that means more light from the same material, and that, again, means more economy. In this direction some improvements have been made, but the progress is hampered by many limitations. Discarding, then, the burner, there remain the three ways first mentioned, which are essentially electrical.

Suppose the light of the immediate future to be a solid rendered incandescent by electricity. Would it not seem that it is better to employ a small button than a frail filament? From many considerations it certainly must be concluded that a button is capable of a higher economy, assuming, of course, the difficulties connected with the operation of such a lamp to be effectively overcome. But to light such a lamp we require a high potential; and to get this economically we must use high frequencies.

Such considerations apply even more to the production of light by the incandescence of a gas, or by phosphorescence. In all cases we require high frequencies and high potentials. These thoughts occurred to me a long time ago.

Incidentally we gain, by the use of very high frequencies, many advantages, such as a higher economy in the light production, the possibility of working with one lead, the possibility of doing away with the leading-in wire, etc.

The question is, how far can we go with frequencies? Ordinary conductors rapidly lose the facility of transmitting electric impulses when the frequency is greatly increased. Assume the means for the production of impulses of very great frequency brought to the utmost perfection, everyone will naturally ask how to transmit them when the necessity arises. In transmitting such impulses through conductors, we must remember that we have to deal with *pressure* and *flow*, in the ordinary interpretation of these terms. Let the pressure increase to an enormous value, and let the flow correspondingly diminish, then such impulses variations merely of pressure, as it were can no doubt be transmitted through a wire even if their frequency be many hundreds of millions per second. It would, of course, be out of question to transmit such impulses through a wire immersed in a gaseous medium, even if the wire were provided with a thick and excellent insulation for most of the energy would be lost in molecular bombardment and consequent heating. The end of the wire connected to the source would be heated, and the remote end would receive but a trifling part of the energy supplied. The prime necessity, then, if such electric impulses are to be used, is to find means to reduce as much as possible the dissipation.

The first thought is, employ the thinnest possible wire surrounded by the thickest practicable insulation. The next thought is to employ electrostatic screens. The insulation of the wire may be covered with a thin conducting coating and the latter connected to the ground. But this would not do, as then all the energy would pass through the conducting coating to the ground, and nothing would get to the end of the wire. If a ground connection is made it can only be made through a conductor offering an enormous impedance, or through a condenser of extremely small capacity. This, however, does not do away with other difficulties.

If the wavelength of the impulses is much smaller than the length of the wire, then corresponding short waves will be sent up in the conducting coating, and it will be more or less the same as though the coating were directly connected to earth. It is therefore necessary to cut up the coating in sections much shorter than the wavelength. Such an arrangement does not still afford a perfect screen, but it is ten thousand times better than none. I think it preferable to cut up the conducting coating in small sections, even if the current waves be much longer than the coating.

If a wire were provided with a perfect electrostatic screen, it would be the same as though all objects were removed from it at infinite distance. The capacity would then be reduced to the capacity of the wire itself, which would be very small. It would then be possible to send over the wire current vibrations of very high frequencies at enormous distance without affecting greatly the character of the vibrations. A perfect screen is of course out of the question, but I believe that with a screen such as I have just described telephony could be rendered practicable across the Atlantic. According to my ideas, the gutta-percha covered wire should be provided with a third conducting coating subdivided in sections. On the top of this should be again placed a layer of gutta-percha and other insulation, and on the top of the whole the armor. But such cables will not be constructed, for ere long intelligence transmitted without wires will throb through the earth like a pulse through a living organism. The wonder is that, with the present state of knowledge and the experiences gained, no attempt is being made to disturb the electrostatic or magnetic condition of the earth, and transmit, if nothing else, intelligence.

It has been my chief aim in presenting these results to point out phenomena or features of novelty, and to advance ideas which I am hopeful will serve as starting points of new departures. It has been my chief desire this evening to entertain you with some novel experiments. Your applause, so frequently and generously accorded, has told me that I have succeeded.

In conclusion, let me thank you most heartily for your kindness and attention, and assure you that the honor I have had in addressing such a distinguished audience, the pleasure I have had in presenting these results to a gathering of so many able men and among them also some of those in whose work for many years past I have found enlightenment and constant pleasure, I shall never forget.

# The Problem Of Increasing Human Energy

# The Onward Movement Of Man – The Energy Of The Movement – The Three Ways Of Increasing Human Energy

Of all the endless variety of phenomena which nature presents to our senses, there is none that fills our minds with greater wonder than that inconceivably complex movement which, in its entirety, we designate as human life; Its mysterious origin is veiled in the forever impenetrable mist of the past, its character is rendered incomprehensible by its infinite intricacy, and its destination is hidden in the unfathomable depths of the future. Whence does it come? What is it? Whither does it end? are the great questions which the sages of all times have endeavoured to answer.

Modern science says: The sun is the past, the earth is the present, the moon is the future. From an incandescent mass we have originated, and into a frozen mass we shall turn. Merciless is the law of nature, and rapidly and irresistibly we are drawn to our doom. Lord Kelvin, in his profound meditations, allows us only a short span of life, something like six million years, after which time the suns bright light will have ceased to shine, and its life-giving heat will have ebbed away, and our own earth will be a lump of ice, hurrying on through the eternal night. But do not let us despair. There will still be left upon it a glimmering spark of life, and there will be a chance to kindle a new fire on some distant star. This wonderful possibility seems, indeed, to exist, judging from Professor Dewar's beautiful experiments with liquid air, which show that germs of organic life are not destroyed by cold, no matter how intense; consequently, they may be transmitted through the interstellar space. Meanwhile the cheering lights of science and art, ever increasing in intensity, illuminate our path, and marvels they disclose, and the enjoyments they offer, make us measurably forgetful of the gloomy future.

Though we may never be able to comprehend human life, we know certainly that it is a movement, of whatever nature it be. The existence of movement unavoidably implies a body which is being moved and a force which is moving it.

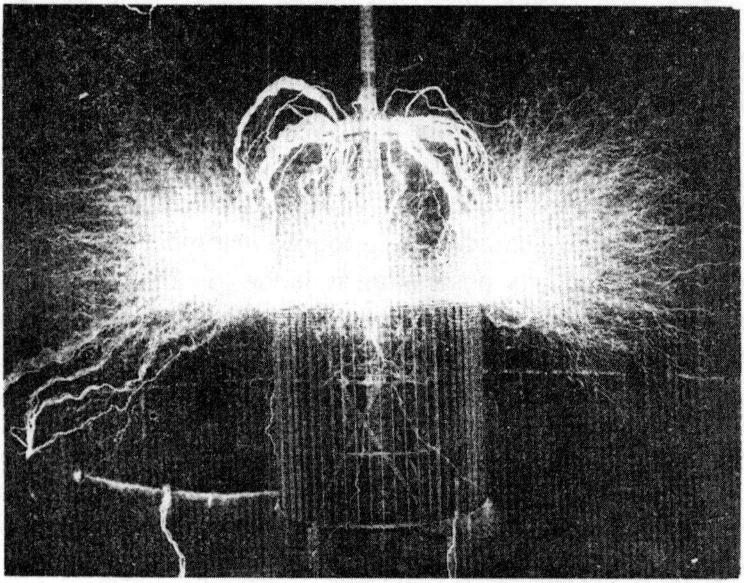

Phot. X. The discharge from the brass ring on the top of "extra coil".

**Fig. 1: Burning the Nitrogen of the atmosphere**

Hence, wherever there is life, there is a mass moved by a force. All mass possesses inertia, all force tends to persist. Owing to this universal property and condition, a body, be it at rest or in motion, tends to remain in the same state, and a force, manifesting itself anywhere and through whatever cause, produces an equivalent opposing force, and as an absolute necessity of this it follows that every movement in nature must be rhythmical. Long ago this simple truth was clearly pointed out by Herbert Spencer, who arrived at it through a somewhat different process of reasoning. It is borne out in everything we perceive in the movement of a planet, in the surging and ebbing of the tide, in the reverberations of the air, the swinging of a pendulum, the oscillations of an electric current, and in the infinitely varied phenomena of organic life. Does not the whole of human life attest to it? Birth, growth, old age, and death of an individual, family, race, or nation, what is it all but a rhythm? All life-manifestation, then, even in its most intricate form, as exemplified in man, how-ever involved and inscrutable, is only a movement, to which the same general laws of movement which govern throughout the physical universe must be applicable.

Note to Fig. 1. This result is produced by the discharge of an electrical oscillator giving twelve million volts. The electrical pressure, alternating one hundred thousand times per second, excites the normally inert nitrogen, causing it to combine with the oxygen. The flame-like discharge shown in the photograph measures sixty-five feet across.

When we speak of man, we have a conception of humanity as a whole, and before applying scientific methods to, the investigation of his movement we must accept this as a physical fact. But can anyone doubt today that all the millions of individuals and all the innumerable types and characters constitute an entity, a unit? Though free to think and act, we are held together, like the stars in the firmament, with ties inseparable. These ties cannot be seen, but we can feel them. I cut myself in the finger, and it pains me: this finger is part of my. I see a friend hurt, and it hurts me, too: my friend and I are one. And now I see stricken down an enemy, a lump of matter which, of all the lumps of matter in the universe, I care least for, and it still grieves me. Does this not prove that each of us is only part of a whole?

For ages this idea has been proclaimed in the consummately wise teachings of religion, probably not alone as a means of insuring peace and harmony among men, but as a deeply founded truth. The Buddhist expresses it in one way, the Christian in another, but both say the same: We are all one. Metaphysical proofs are, however, not the only ones which we are able to bring forth in support of this idea. Science, too, recognizes this connected ness of separate individuals, though not quite in the same sense as it admits that the suns, planets, and moons of a constellation are one body, and there can be no doubt that it will be experimentally confirmed in times to come, when our means and methods for investigating psychical and other states and phenomena shall have been brought to great perfection. Still more: this one human being lives on and on. The individual is ephemeral, races and nations come and pass away, but man remains. Therein lies the profound difference between the individual and the whole. Therein, too, is to be found the partial explanation of many of those marvellous phenomena of heredity which are the result of countless centuries of feeble but persistent influence.

Conceive, then, man as a mass urged on by a force. Though this movement is not of a translatory character, implying change of place, yet the general laws of mechanical movement are applicable to it, and the energy associated with this mass can be measured, in accordance with well-known principles, by half the product of the mass with the square of a certain velocity. So, for instance, a can-non-ball which is at rest possesses a certain amount of energy in the form of heat, which we measure in a similar way. We imagine the ball to consist of innumerable minute particles, called atoms or molecules, which vibrate or whirl around one another. We determine their masses and velocities, and from them the energy of each of these minute systems, and adding them all together, we get an idea of the total heat-energy contained in the ball, which is only seemingly at rest. In this purely theoretical estimate this energy may then be calculated by multiplying half of the total mass that is half of the sum of all the small masses--with the square of a velocity which is determined from the velocities of the separate particles. In like manner we may conceive of human energy being measured by half the human mass multiplied with the square of the velocity which we are not yet able to compute. But our deficiency in this knowledge will not vitiate the truth of the deductions I shall draw, which rest on the firm basis that the same laws of mass and force govern throughout nature.

Man, however, is not an ordinary mass, consisting of spinning atoms and molecules, and containing merely heat-energy. He is a mass possessed of certain higher qualities by reason of the creative principle of life with which he is endowed. His mass, as the water in an ocean wave, is being continuously exchanged, new taking the place of the old. Not only this, but he grows propagates, and dies, thus altering his mass independently, both in bulk and density. What is most wonderful of all, he is capable of increasing or diminishing his velocity of movement by the mysterious power he possesses by appropriating more or less energy from other substance and turning it into motive energy. But in any given moment we may ignore these slow changes and assume that human energy is measured by half the product of man's mass with the square of a certain hypothetical velocity. How-ever we may compute this velocity, and whatever we may take as the standard of its measure, we must, in harmony with this conception, come to the conclusion that the great problem of science is, and always will be, to increase the energy thus defined.

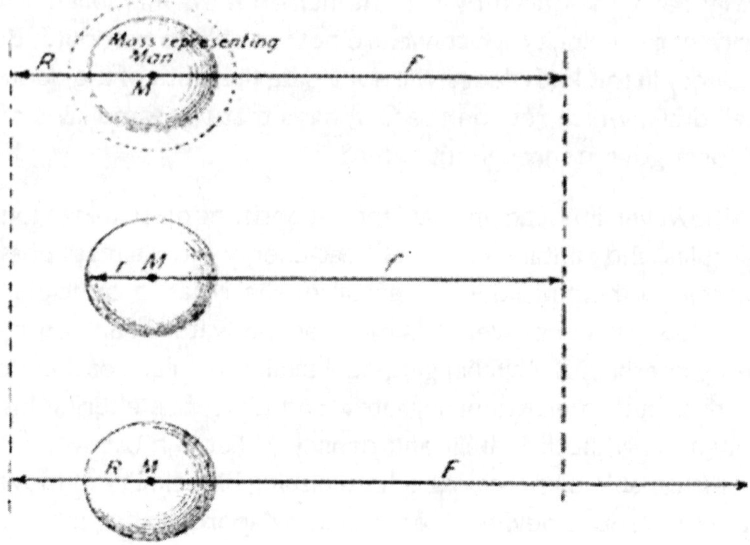


**DIAGRAM a: The three ways of increasing human energy.**

Many years ago, stimulated by the perusal of that deeply interesting work, Draper's "History of the Intellectual Development of Europe," depicting so vividly human movement, I recognized that to solve this eternal problem must ever be the chief task of the man of science. Some results of my own efforts to this end I shall endeavour briefly to describe here.

Let, then, in diagram a, M represent the mass of man. This mass is impelled in one direction by a force f, which is resisted by another partly frictional and partly negative force R, acting in a direction exactly opposite, and retarding the movement of the mass. Such an antagonistic force is present in every movement and must be taken into consideration. The difference between these two forces is the effective force which imparts a velocity V to the mass M in the direction of the arrow on the line representing the force f. In accordance with the preceding, the human energy will then be given by the product ½ MV2 = ½ MV x V, in which M is the total mass of man in the ordinary interpretation of the term "mass," and V is a certain hypothetical velocity, which, in the present state of science, we are unable exactly to define and determine. To increase the human energy is, there-fore, equivalent to increasing this product, and there are, as will readily be seen, only three ways possible to attain this result, which are illustrated in the above diagram. The first way shown in the top figure, is to increase the mass (as indicated by the dotted circle), leaving the two opposing forces the same. The second way is to reduce the retarding force R to a smaller value r, leaving the mass and the impelling force the same, as diagrammatically shown in the middle figure. The third way, which is illustrated in the last figure, is to increase the impelling force f to a higher value F, while the mass and the retarding force R remain unaltered. Evidently fixed limits exist as regards increase of mass and reduction of retarding force, but the impelling force can be increased indefinitely. Each of these three possible solutions presents a different aspect of the main problem of increasing human energy, which is thus divided into three dis-tinct problems, to be successively considered.

# The First Problem: How To Increase The Human Mass – The Burning Of Atmospheric Nitrogen

Viewed generally, there are obviously two ways of increasing the mass of mankind: first, by aiding and maintaining those forces and conditions which tend to increase it and second, by opposing and reducing those which tend to diminish it. The mass will be increased by careful attention to health, by substantial food, by moderation, by regularity of habits, by promotion of marriage, by conscientious attention to children, and, generally stated, by the observance of all the many precepts and laws of religion and hygiene. But in adding new mass to the old, three cases again present themselves. Either the mass added is of the same velocity as the old, or it is of a smaller or of a higher velocity. To gain an idea of the relative importance of these cases, imagine a train com-posed of, say, one hundred locomotives running on a track, and suppose that, to increase the energy of the moving mass, four more locomotives are added to the train. If these four moves at the same velocity at which the train is going, the total energy will be increased four per cent.; if they are moving at only one half of that velocity, the increase will amount to only one per cent.; if they are moving at twice that velocity, the increase of energy will be sixteen per cent. This simple illustration shows that it is of greatest importance to add mass of a higher velocity. Stated more to the point, if, for example, the children be of the same degree of enlightenment as the parents, that is, mass of the "same velocity," the energy will simply increase proportionately to the number added. If they are less intelligent or advanced, or mass of "smaller velocity," there will be a very slight gain in the energy; but if they are further advanced, or mass of "higher velocity," then the new generation will add very considerably to the sum total of human energy. any addition of mass of "smaller velocity," beyond that indispensable amount required by the law expressed in the proverb, "Mens sana in corpore sano," should be strenuously opposed. For instance, the mere development of muscle, as aimed at in some of our colleges, I consider equivalent to adding mass of "smaller velocity," and I would not commend it, although my views were different when I was a student myself. Moderate exercise, insuring the right balance between mind and body, and the highest efficiency of performance, is, of course, a prime requirement. The above example shows that the most important result to be attained is the education, or the increase of the "velocity," of the mass newly added.

Conversely, it scarcely need be stated that everything that is against the teachings of religion and the laws, of hygiene is tending to decrease the mass. Whisky, wine, tea coffee, tobacco, and other such stimulants are responsible for the shortening of the lives of many, and ought to be used with moderation. But I do not think that rigorous measures of suppression of habits followed through many generations are commendable. It is wiser to preach moderation than abstinence. We have become accustomed to these stimulants, and if such reforms are to be effected, they must be slow and gradual. Those who are devoting their energies to such ends could make themselves far more useful by turning their efforts in other directions, as, for instance, toward providing pure water.

For every person who perishes from the effects of a stimulant, at least a thousand die from the consequences of drinking impure water. This precious fluid, which daily infuses new life into us, is likewise the chief vehicle through which disease and death enter our bodies. The germs of destruction it conveys are enemies all the more terrible as they perform their fatal work unperceived. They seal our doom while we live and enjoy. The majority of people are so ignorant or careless in drinking water, and the consequences of this are so disastrous, that a philanthropist can scarcely use his efforts better than by endeavouring to enlighten those who are thus injuring themselves. By systematic purification and sterilization of the drinking water the human mass would be very considerably increased. It should be made a rigid rule which might be enforced by law to boil or to sterilize otherwise the drinking water in every household and public place. The mere filtering does not afford sufficient security against infection. All ice for internal uses should be artificially prepared from water thoroughly sterilized. The importance of eliminating germs of disease from the city water is generally recognized, but little is being done to improve the existing conditions, as no satisfactory method of sterilizing great quantities of water has yet been brought forward. By improved electrical appliances we are now enabled to produce ozone cheaply and in large amounts, and this ideal disinfectant seems to offer a happy solution of the important question.

Gambling, business rush, and excitement, particularly on the exchanges, are causes of much mass reduction, all the more so because the individuals concerned represent units of higher value. Incapacity of observing the first symptoms of an illness, and careless neglect of the same, are important factors of mortality. In noting carefully every new sign of approaching danger, and making conscientiously every possible effort to avert it, we are not only following wise laws of hygiene in the interest of our well-being and the success of our labours, but we are also complying with a higher moral duty. Everyone should consider his body as a priceless gift from one whom he loves above all, as a marvellous work of art, of indescribable beauty and mastery beyond human conception, and so delicate and frail that a word, a breath, a look, nay, a thought, may injure it. Uncleanness, which breeds disease and death, is not only a self-destructive but highly immoral habit. In keeping our bodies free from infection, healthful, and pure, we are expressing our reverence for the high principle with which they are endowed. He who follows the precepts of hygiene in this spirit is proving him-self, so far, truly religious. Laxity of morals is a terrible evil, which poisons both mind and body, and which is responsible for a great reduction of the human mass in some countries. Many of the present customs and tendencies are productive of similar hurtful results. For example, the society life, modern education, and pursuits of women, tending to draw them away from their household duties and make men out of them, must needs detract from the elevating ideal they represent, diminish the artistic creative power, and cause sterility and a general weakening of the race. A thousand other evils might be mentioned, but all put together, in their bearing upon the problem under discussion, they could not equal a single one, the want of food, brought on by poverty, destitution, and famine. Millions of individuals die yearly for want of food, thus keeping down the mass. Even in our enlightened communities, and notwithstanding the many charitable efforts, this is still, in all probability, the chief evil. I do not mean here absolute want of food but want of healthful nutriment.

How to provide good and plentiful food is, therefore, a most important question of the day. On the general principles the raising of cattle as a means of providing food is objectionable, because, in the sense interpreted above, it must undoubtedly tend to the addition of mass of a "smaller velocity." It is certainly preferable to raise vegetables, and I think, therefore, that vegetarianism is a commend-able departure from the established barbarous habit. That we can subsist on plant food and per-form our work even to advantage is not a theory, but a well-demonstrated fact. Many races living almost exclusively on vegetables are of superior physique and strength. There is no doubt that some plant food, such as oatmeal, is more economical than meat, and superior to it in regard to both mechanical and mental performance. Such food, moreover, taxes our digestive organs decidedly less, and, in making us more contented and sociable, produces an amount of good difficult to estimate. In view of these facts every effort should be made to stop the wanton and cruel slaughter of animals, which must be destructive to our morals. To free ourselves from animal instincts and appetites, which keep us down, we should begin at the very root from which we spring: we should effect a radical reform in the character of the food.

There seems to be no *philosophical* necessity for food. We can conceive of organized beings living without nourishment and deriving all the energy they need for the performance of their life functions from the ambient medium. In a crystal we have the clear evidence of the existence of a formative life-principle, and though we cannot understand the life of a crystal, it is none the less a living being. There may be, besides crystals, other such individualized, material systems of beings, perhaps of gaseous constitution, or composed of substance still more tenuous. In view of this possibility, nay, probability, we cannot apodictically deny the existence of organized beings on a planet merely because the conditions on the same are unsuitable for the existence of life as we conceive it. We cannot even, with positive assurance, assert that some of them might not be present here, in this our world, in the very midst of us, for their constitution and life-manifestation may be such that we are unable to perceive them.

The production of artificial food as a means for causing an increase of the human mass naturally suggests itself, but a direct attempt of this kind to provide nourishment does not appear to me rational, at least not for the present. Whether we could thrive on such food is very doubtful. We are the result of ages of continuous adaptation, and we cannot radically change without unforeseen and, in all probability, disastrous consequences. So un-certain an experiment should not be tried. By far the best way, it seems to me, to meet the ravages of the evil, would be to find ways of increasing the productivity of the soil. With this object the preservation of forests is of an importance which cannot be overestimated, and in this connection, also, the utilization of waterpower for purposes of electrical transmission, dispensing in many ways with the necessity of burning wood, and tending thereby to forest preservation, is to be strongly advocated. But there are limits in the improvement to be effected in this and similar ways.

To increase materially the productivity of the soil, it must be more effectively fertilized by artificial means. The question of food-production resolves itself, then, into the question how best to fertilize the soil. What it is that made the soil is still a mystery. To explain its origin is probably equivalent to explaining the origin of life itself. The rocks, disintegrated by moisture and heat and wind and weather, were in themselves not capable of maintaining life. Some unexplained condition arose, and some new principle came into effect, and the first layer capable of sustaining low organisms, like mosses was formed. These, by their life and death, added more of the life sustaining quality to the soil, and higher organisms could then subsist, and so on and on, until at last highly developed plant and animal life could flourish. But though the theories are, even now, not in agreement as to how fertilization is effected, it is a fact, only too well ascertained, that the soil cannot indefinitely sustain life, and some way must be found to supply it with the substances which have been abstracted from it by the plants. The chief and most valuable among these substances are compounds of nitrogen, and the cheap production of these is, therefore, the key for the solution of the all-important food problem. Our atmosphere contains an inexhaustible amount of nitrogen and could we but oxidize it and produce these compounds, an incalculable benefit for mankind would follow.

Long ago this idea took a powerful hold on the imagination of scientific men, but an efficient means for accomplishing this result could not be devised. The problem was rendered extremely difficult by the extraordinary inertness of the nitrogen, which refuses to combine even with oxygen. But here electricity comes to our aid: the dormant affinities of the element are awakened by an electric current of the proper quality. As a lump of coal which has been in contact with oxygen for centuries without burning will combine with it when once ignited, so nitrogen, excited by electricity, will burn. I did not succeed, however, in producing electrical discharges exciting very effectively the atmospheric nitrogen until a comparatively recent date, although I showed, in May 1891, in a scientific lecture, a novel form of discharge or electrical flame named "St. Elmo's hot-fire," which, besides being capable of generating ozone in abundance, also possessed, as I pointed out on that occasion, distinctly the quality of exciting chemical affinities. This discharge or flame was then only three or four inches long, its chemical action was likewise very feeble, and consequently the process of oxidation of nitrogen was wasteful. How to intensify this action was the question. Evidently electric currents of a peculiar kind had to be produced in order to render the process of nitrogen combustion more efficient.

The first advance was made in ascertaining that the chemical activity of the discharge was very considerably increased by using currents of extremely high frequency or rate of vibration. This was an important improvement, but practical considerations soon set a definite limit to the progress in this direction. Next, the effects of the electrical pressure of the current impulses, of their wave form and other characteristic features, were investigated. Then the influence of the atmospheric pressure and temperature and of the presence of water and other bodies was studied, and thus the best conditions for causing the most intense chemical action of the discharge and securing the highest efficiency of the process were gradually ascertained. Naturally, the improvements were not quick in coming; still, little by little, I advanced. The flame grew larger and larger, and its oxidizing action grew more intense. From an insignificant brush-discharge a few inches long it developed into a marvellous electrical phenomenon, a roaring blaze, devouring the nitrogen of the atmosphere and measuring sixty or seventy feet across. Thus slowly, almost imperceptibly, possibility became accomplishment. All is not yet done, by any means, but to what a degree my efforts have been rewarded an idea may be gained from an inspection of Fig. 1 (p. 176), which, with its title, is self-explanatory. The flame-like discharge visible is produced by the intense electrical oscillations which pass through the coil shown, and violently agitate the electrified molecules of the air. By this means a strong affinity is created between the two normally indifferent constituents of the atmosphere, and they combine readily, even if no further provision is made for intensifying the chemical action of the discharge. In the manufacture of nitrogen compounds by this method, of course, every possible means bearing upon the intensity of this action and the efficiency of the process will be taken advantage of, and, besides, special arrangements will be provided for the fixation of the compounds formed, as they are generally unstable, the nitrogen becoming again inert after a little lapse of time. Steam is a simple and effective means for fixing permanently the compounds. The result illustrated makes it practicable to oxidize the atmospheric nitrogen in unlimited quantities, merely by the use of cheap mechanical power and simple electrical apparatus. In this manner many compounds of nitrogen may be manufactured all over the world, at a small cost, and in any desired amount, and by means of these compounds the soil can be fertilized and its productiveness indefinitely increased. An abundance of cheap and healthful food, not artificial, but such as we are accustomed to, may thus be obtained. This new and inexhaustible source of food-supply will be of incalculable benefit to mankind, for it will enormously contribute to the increase of the human mass, and thus add immensely to human energy. Soon, I hope, the world will see the beginning of an industry which, in time to come, will, I believe, be in importance next to that if iron.

# The Second Problem: How To Reduce The Force Retarding The Human Mass – The Art Of Telautomatics

As before stated, the force which retards the onward movement of man is partly frictional and partly negative. To illustrate this distinction I may name, for example, ignorance, stupidity, and imbecility as some of the purely frictional forces, or resistances devoid of any directive tendency. On the other hand, visionariness, insanity, self-destructive tendency, religious fanaticism, and the like, are all forces of a negative character, acting in definite directions. To reduce or entirely overcome these dissimilar retarding forces, radically different methods must be employed. One knows, for in-stance, what a fanatic may do, and one can take preventive measures, can enlighten, convince, and possibly direct him, turn his vice into virtue; but one does not know, and never can know, what a brute or an imbecile may do, and one must deal with him as with a mass, inert, without mind, let loose by the mad elements. A negative force al-ways implies some quality, not infrequently a high one, though badly directed, which it is possible to turn to good advantage; but a directionless, frictional force involves unavoidable loss. Evidently, then, the first and general answer to the above question is: turn all negative force in the right direction and reduce all frictional force.

There can be no doubt that, of all the frictional resistances, the one that most retards human movement is ignorance. Not without reason said that man of wisdom, Buddha: "Ignorance is the greatest evil in the world." The friction which results from ignorance, and which is greatly in-creased owing to the numerous languages and nationalities, can be reduced only by the spread of knowledge and the unification of the heterogeneous elements of humanity. No effort could be better spent. But however, ignorance may have retarded the onward movement of man in times past, it is certain that, nowadays, negative forces have become of greater importance. Among these there is one of far greater moment than any other. It is called organized warfare. When we consider the millions of individuals, often the ablest in mind and body, the flower of humanity, who are compelled to a life of inactivity and unproductiveness, the immense sums of money daily required for the maintenance of armies and war apparatus, representing ever so much of human energy, all the effort uselessly spent in the production of arms and implements of destruction, the loss of life and the fostering of a barbarous spirit, we are appalled at the inestimable loss to mankind which the existence of these deplorable conditions must involve. What can we do to combat best this great evil?

Law and order absolutely require the maintenance of organized force. No community can exist and prosper without rigid discipline. Every country must be able to defend itself, should the necessity arise. The conditions of to-day are not the result of yesterday, and a radical change cannot be effected to-morrow. If the nations would at once disarm, it is more than likely that a state of things worse than war itself would follow. Universal peace is a beautiful dream, but not at once realizable. We have seen recently that even the noble effort of the man invested with the greatest worldly power has been virtually without effect. And no wonder, for the establishment of universal peace is, for the time being, a physical impossibility. War is a negative force and cannot be turned in a positive direction without passing through, the intermediate phases. It is a problem of making a wheel, rotating one way, turn in the opposite direction without slowing it down, stopping it, and speeding it up again the other way.

It has been argued that the perfection of guns of great destructive power will stop warfare. So, I myself thought for a long time, but now I believe this to be a profound mistake. Such developments will greatly modify, but not arrest it. On the contrary, I think that every new arm that is invented, every new departure that is made in this direction, merely invites new talent and skill, engages new effort, offers new incentive, and so only gives a fresh impetus to further development. Think of the discovery of gunpowder. Can we conceive of any more radical departure than was effected by this innovation? Let us imagine ourselves living in that period: would we not have thought then that warfare was at an end, when the armor of the knight became an object of ridicule, when bodily strength and skill, meaning so much before, became of comparatively little value? Yet gunpowder did not stop warfare: quite the opposite it acted as a most powerful incentive. Nor do I believe that warfare can ever be arrested by any scientific or ideal development, so long as similar conditions to those prevailing now exist, because war has itself be-come a science, and because war involves some of the most sacred sentiments of which man is capable. In fact, it is doubtful whether men who would not be ready to fight for a high principle would be good for anything at all. It is not the mind which makes man, nor is it the body; it is mind and body. Our virtues and our failings are inseparable, like force and matter. When they separate, man is no more.

Another argument, which carries considerable force, is frequently made, namely, that war must soon become impossible because the means of defence are outstripping the means of attack. This is only in accordance with a fundamental law which may be expressed by the statement that it is easier to destroy than to build. This law defines human capacities and human conditions. Were these such that it would be easier build than to destroy, man would go on unresisted, creating and accumulating without limit. Such conditions are not of this earth. A being which could do this would not be a man: it might be a god. Défense will always have the advantage over attack, but this alone, it seems to me, can never stop war. By the use of new principles of defence, we can render harbours impregnable against attack, but we cannot by such means prevent two warships meeting in battle on the high sea. And then, if we follow this idea to its ultimate development, we are led to the conclusion that it would be better for mankind if attack and defence were just oppositely related; for if every country, even the smallest, could surround itself with a wall absolutely impenetrable, and could defy the rest of the world, a state of things would surely be brought on which would be extremely unfavourable to human progress. It is by abolishing all the barriers which separate nations and countries that civilization is best furthered.

Again, it is contended by some that the advent of the flying-machine must bring on universal peace. This, too, I believe to be an entirely erroneous view. The flying-machine is certainly coming, and very soon, but the conditions will remain the same as before. In fact, I see no reason why a ruling power, like Great Britain, might not govern the air as well as the sea. Without wishing to put myself on record as a prophet, I do not hesitate to say that the next years will see the establishment of an "air-power," and its centre may be not far from New York. But, for all that, men will fight on merrily.

The ideal development of the war principle would ultimately lead to the transformation of the whole energy of war into purely potential, explosive energy, like that of an electrical condenser. In this form the war-energy could be maintained without effort; it would need to be much smaller in amount, while incomparably more effective.

As regards the security of a country against foreign invasion, it is interesting to note that it depends only on the relative, and not the absolute, number of the individuals or magnitude of the forces, and that, if every country should reduce the war-force in the same ratio, the security would remain unaltered. An international agreement with the object of reducing to a minimum the war-force which, in view of the present still imperfect education of the masses, is absolutely indispensable, would, therefore, seem to be the first rational step to take toward diminishing the force retarding human movement.

Fortunately, the existing conditions cannot continue indefinitely, for a new element is beginning to assert itself. A change for the better is eminent, and I shall now endeavour to show what, according to my ideas, will be the first advance toward the establishment of peaceful relations between nations, and by what means it will eventually be accomplished.

Let us go back to the early beginning, when the law of the stronger was the only law. The light of reason was not yet kindled, and the weak was entirely at the mercy of the strong. The weak individual then began to learn how to defend himself. He made use of a club, stone, spear, sling, or bow and arrow, and in the course of time, instead of physical strength, intelligence became the chief deciding factor in the battle. The wild char-acter was gradually softened by the awakening of noble sentiments, and so, imperceptibly, after ages of continued progress, we have come from the brutal fight of the unreasoning animal to what we call the "civilized warfare" of to-day, in which the combatants shake hands, talk in a friendly way, and smoke cigars in the entr'actes, ready to engage again in deadly conflict at a signal. Let pessimists say what they like, here is an absolute evidence of great and gratifying advance.

But now, what is the next phase in this evolution? Not peace as yet, by any means. The next change which should naturally follow from modern developments should be the continuous diminution of the number of individuals engaged in battle. The apparatus will be one of specifically great power, but only a few individuals will be required to operate it. This evolution will bring more and more into prominence a machine or mechanism with the fewest individuals as an element of warfare, and the absolutely unavoidable consequence of this will be the abandonment of large, clumsy, slowly moving, and unmanageable units. Greatest possible speed and maximum rate of energy-delivery by the war apparatus will be the main object. The loss of life will become smaller and smaller, and finally, the number of the individuals continuously diminishing, merely machines will meet in a contest without bloodshed, the nation's being simply interested, ambitious spectators. When this happy condition is realized, peace will be assured. But no matter to what degree of perfection rapid-fire guns, high-power cannon, explosive projectiles, torpedo-boats, or other implements of war may be brought, no matter how destructive they may be made, that condition can never be reached through any such development. All such implements require men for their operation; men are indispensable parts of the machinery. Their object is to kill and to destroy. Their power resides in their capacity for doing evil. So long as men meet in battle, there will be blood-shed. Bloodshed will ever keep up barbarous passion. To break this fierce spirit, a radical departure must be made, an entirely new principle must be introduced, something that never existed before in warfare a principle which will forcibly, unavoidably, turn the battle into a mere spectacle, a play, a contest without loss of blood. To bring on this result men must be dispensed with machine must fight machine. But how accomplish that which seems impossible? The answer is simple enough: produce a machine capable of acting as though it were part of a human being--no mere mechanical contrivance, comprising levers, screws, wheels, clutches, and nothing more, but a machine embodying a higher principle, which will enable it to perform its duties as though it had intelligence, experience, judgment, a mind! This conclusion is the result of my thoughts and observations which have extended through virtually my whole life, and I shall now briefly describe how I came to accomplish that which at first seemed an unrealizable dream.

A long time ago, when I was a boy, I was afflicted with a singular trouble, which seems to have been due to an extraordinary excitability of the retina. It was the appearance of images which, by their persistence, marred the vision of real objects and interfered with thought. When a word was said to me, the image of the object which it designated would appear vividly before my eyes, and many times it was impossible for me to tell whether the object I saw was real or not. This caused me great discomfort and anxiety, and I tried hard to free myself of the spell. But for a long time, I tried in vain, and it was not, as I clearly recollect, until I was about twelve years old that I succeeded for the first time, by an effort of the will, in banishing an image which presented itself. My happiness will never be as complete as it was then, but unfortunately (as I thought at that time), the old trouble returned, and with it my anxiety. Here it was that the observations to which I refer began. I noted, namely, that whenever the image of an object appeared before my eyes, I had seen some-thing that reminded me of it. In the first instances I thought this to be purely accidental, but soon I convinced myself that it was not so. A visual impression, consciously or unconsciously received, invariably preceded the appearance of the image. Gradually the desire arose in me to find out, every time, what caused the images to appear, and the satisfaction of this desire soon became a necessity. The next observation I made was that, just as these images followed as a result of something I had seen, so also the thoughts which I conceived were suggested in like manner. Again, I experienced the same desire to locate the image which caused the thought, and this search for the original visual impression soon grew to be a second nature. Mt mind became automatic, as it were, and in the course of years of continued, almost unconscious performance, I acquired the ability of locating every time and, as a rule, instantly the visual impression which started the thought. Nor is this all. It was not long before I was aware that also all my movements were prompted in the same way, and so, searching, observing, and verifying continuously, year by year, I have, by every thought and every act of mine, demonstrated, and do so daily, to my absolute satisfaction, that I am an automaton endowed with power of movement, which merely responds to external stimuli beating upon my sense organs, and thinks and acts and moves accordingly. I remember only one or two cases in all my life in which I was unable to locate the first impression which prompted a movement or a thought, or even a dream.

Fig. 2:The first practical Tele automation.

A machine having all the bodily or translatory movements and the operations of the interior mechanism controlled from a distance without wires. The crewless boat shown in the photograph contains its own motive power, propelling and steering machinery, and numerous other accessories, all of which are controlled by transmitting from a distance, without wires, electrical oscillations to a circuit carried by the boat and adjusted to respond only to these oscillations.

With these experiences it was only natural that, long ago, I conceived the idea of constructing an automaton which would mechanically represent me, and which would respond, as I do myself, but, of course, in a much more primitive manner, to external influences. Such an automaton evidently had to have motive power, organs for locomotion, directive organs, and one or more sensitive organs so adapted as to be excited by external stimuli. This machine would, I reasoned, perform its movements in the manner of a living being, for it would have all the chief mechanical characteristics or elements of the same. There was still the capacity for growth, propagation, and, above all, the mind which would be wanting to make the model complete. But growth was not necessary in this case, since a machine could be manufactured full-grown, so to speak. As to the capacity for propagation, it could likewise be left out of consideration, for in the mechanical model it merely signified a process of manufacture. Whether the automation be of flesh and bone, or of wood and steel, it mattered little, provided it could perform all the duties re-quired of it like an intelligent being. To do so, it had to have an element corresponding to the mind, which would affect the control of all its movements and operations, and cause it to act, in any unforeseen case that might present itself, with knowledge, reason, judgment, and experience. But this element I could easily embody in it by conveying to it my own intelligence, my own under-standing. So, this invention was evolved, and so a new art came into existence, for which the name "tele automatics" has been suggested, which means the art of controlling the movements and operations of distant automatons. This principle evidently was applicable to any kind of machine that moves on land or in the water or in the air. In applying it practically for the first time, I selected a boat (see Fig. 2). A storage battery placed within it furnished the motive power. The propeller, driven by a motor, represented the locomotive organs. The rudder, controlled by another motor likewise driven by the battery, took the place of the directive organs. As to the sensitive organ, obviously the first thought was to utilize a device responsive to rays of light, like a selenium cell, to represent the human eye. But upon closer inquiry I found that, owing to experimental and other difficulties, no thoroughly satisfactory control of the automaton could be effected by light, radiant heat, Hertzian radiations, or by rays in general, that is, disturbances which pass in straight lines through space. One of the reasons was that any obstacle coming between the operator and the distant automaton would place it beyond his control. Another reason was that the sensitive device representing the eye would have to be in a definite position with respect to the distant controlling apparatus, and this necessity would impose great limitations in the control. Still another and very important reason was that, in using rays, it would be difficult, if not impossible, to give to the automaton individual features or characteristics distinguishing it from other machines of this kind. Evidently the automaton should respond only to an individual call, as a person responds to a name. Such considerations led me to conclude that the sensitive device of the machine should correspond to the ear rather than the eye of a human being, for in this case its actions could be controlled irrespective of intervening obstacles, regardless of its position relative to the distant controlling apparatus, and, last, but not least, it would remain deaf and unresponsive, like a faithful servant, to all calls but that of its master. These requirements made it imperative to use, in the control of the automaton, instead of light or other rays, waves or disturbances which propagate in all directions through space, like sound, or which follow a path of least resistance, however curved. I attained the result aimed at by means of an electric circuit placed within the boat, and adjusted, or "tuned," exactly to electrical vibrations of the proper kind transmitted to it from a distant "electrical oscillator." This circuit, in responding, how-ever feebly, to the transmitted vibrations, affected magnets and other contrivances, through the medium of which were controlled the movements of the propeller and rudder, and also the operations of numerous other appliances.

By the simple means described the knowledge, experience, judgment--the mind, so to speak--of the distant operator were embodied in that machine, which was thus enabled to move and to perform all its operations with reason and intelligence. It behaved just like a blindfolded person obeying directions received through the ear.

The automatons so far constructed had "borrowed minds," so to speak, as each merely formed part of the distant operator who conveyed to it his intelligent orders; but this art is only in the beginning. I purpose to show that, however impossible it may now seem, an automaton may be contrived which will have its "own mind," and by this, I mean that it will be able, independent of any operator, left entirely to itself, to perform, in response to external influences affecting its sensitive organs, a great variety of acts and operations as if it had intelligence. It will be able to follow a course laid out or to obey orders given far in advance; it will be capable of distinguishing between what it ought and what it ought not to do, and of making experiences or otherwise stated, of recording impressions which will definitely affect its subsequent actions. In fact, I have already conceived such a plan.

Although I evolved this invention many years ago and explained it to my visitors very frequently in my laboratory demonstrations, it was not until much later, long after I had perfected it, that it became known, when, naturally enough, it gave rise to much discussion and to sensational reports. But the true significance of this new art was not grasped by the majority, nor was the great force of the underlying principle recognized. As nearly as I could judge from the numerous comments which appeared, the results I had obtained were considered as entirely impossible. Even the few who were disposed to admit the practicability of the invention saw in it merely an automobile torpedo, which was to be used for the purpose of blowing up battleships, with doubtful success. The general impression was that I contemplated simply the steering of such a vessel by means of Hertzian or other rays. There are torpedoes steered electrically by wires, and there are means of communicating without wires, and the above was, of course an obvious inference. Had I accomplished nothing more than this, I should have made a small advance indeed. But the art I have evolved does not contemplate merely the change of direction of a moving vessel; it affords means of absolutely controlling, in every respect, all the innumerable translatory movements, as well as the operations of all the internal organs, no matter how many, of an individualized automaton. Criticisms to the effect that the control of the automaton could be interfered with were made by people who do not even dream of the wonderful results which can be accomplished by use of electrical vibrations. The world moves slowly, and new truths are difficult to see. Certainly, by the use of this principle, an arm for attack as well as defence may be provided, of a destructiveness all the greater as the principle is applicable to submarine and aerial vessels. There is virtually no restriction as to the amount of explosive it can carry, or as to the distance at which it can strike, and failure is almost impossible. But the force of this new principle does not wholly reside in its destructiveness. Its advent introduces into warfare an element which never existed before a fighting machine without men as a means of attack and defence. The continuous development in this direction must ultimately make war a mere contest of machines without men and without loss of life, a condition which would have been impossible without this new departure, and which, in my opinion, must be reached as preliminary to permanent peace. The future will either bear out or disprove these views. My ideas on this subject have been put forth with deep conviction, but in a humble spirit.

The establishment of permanent peaceful relations between nations would most effectively reduce the force retarding the human mass and would be the best solution of this great human problem. But will the dream of universal peace ever be realized? Let us hope that it will. When all darkness shall be dissipated by the light of science, when all nations shall be merged into one, and patriotism shall be identical with religion, when there shall be one language, one country, one end, then the dream will have become reality.

# The Third Problem: How To Increase The Force Accelerating The Human Mass —The Harnessing Of The Sun's Energy

Of the three possible solutions of the main problem of increasing human energy, this is by far the most important to consider, not only because of its intrinsic significance, but also because of its intimate bearing on all the many elements and conditions which determine the movement of humanity. In order to proceed systematically, it would be necessary for me to dwell on all those considerations which have guided me from the outset in my efforts to arrive at a solution, and which have led me, step by step, to the results I shall now describe. As a preliminary study of the problem an analytical investigation, such as I have made, of the chief forces which determine the onward movement, would be of advantage, particularly in conveying an idea of that hypothetical "velocity" which, as explained in the beginning, is a measure of human energy; but to deal with this specifically here, as I would desire, would lead me far beyond the scope of the present subject. Suffice it to state that the resultant of all these forces is always in the direction of reason, which therefore, determines, at any time, the direction of human movement. This is to say that every effort which is scientifically applied, rational, useful, or practical, must be in the direction in which the mass is moving. The practical, rational man, the observer, the man of business, he who reasons, calculates, or determines in advance, carefully applies his effort so that when coming into effect it will be in the direction of the movement, making it thus most efficient, and in this knowledge and ability lies the secret of his success. Every new fact discovered, every new experience or new element added to our knowledge and entering into the domain of reason, affects the same and, therefore, changes the direction of movement, which, however, must always take place along the resultant of all those efforts which, at that time, we designate as reason-able, that is, self-preserving, useful, profitable, or practical. These efforts concern our daily life, our necessities and comforts, our work and business, and it is these which drive man onward.

But looking at all this busy world about us, on all this complex mass as it daily throbs and moves, what is it, but an immense clockwork driven by a spring? In the morning, when we rise, we cannot fail to note that all the objects about us are manufactured by machinery: the water we use is lifted by steam-power; the trains bring our breakfast from distant localities; the elevators in our dwelling and our office building, the cars that carry us there, are all driven by power; in all our daily errands, and in our very life-pursuit, we depend upon it; all the objects we see tell us of it; and when we return to our machine-made dwelling at night, lest we should forget it, all the material comforts of our home, our cheering stove and lamp, remind us of how much we depend on power. And when there is an accidental stoppage of the machinery, when the city is snowbound, or the life sustaining movement otherwise temporarily arrested, we are affrighted to realize how im-possible it would be for us to live the life we live without motive power. Motive power means work. To increase the force accelerating human movement means, therefore, to perform more work.

So, we find that the three possible solutions of the great problem of increasing human energy are answered by the three words: *food, peace, work*. Many a year I have thought and pondered, lost myself in speculations and theories, considering man as a mass moved by a force, viewing his inexplicable movement in the light of a mechanical one, and applying the simple principles of mechanics to the analysis of the same until I arrived at these solutions, only to realize that they were taught to me in my early childhood. These three words sound the keynotes of the Christian religion. Their scientific meaning and purpose now clear to me: food to increase the mass, peace to diminish the retarding force, and work to increase the force accelerating human movement. These are the only three solutions which are possible of that great problem, and all of them have one object, one end, namely, to increase human energy. When we recognize this, we cannot help wondering how profoundly wise and scientific and how immensely practical the Christian religion is, and in what a marked contrast it stands in this respect to other religions. It is unmistakably the result of practical experiment and scientific observation which have extended through the ages, while other religions seem to be the outcome of merely abstract reasoning. Work, untiring effort, useful and accumulative, with periods of rest and recuperation aiming at higher efficiency, is its chief and ever-recurring command. Thus, we are inspired both by Christianity and Science to do our utmost toward increasing the performance of mankind. This most important of human problems I shall now specifically consider.

# The Source Of Human Energy – The Three Ways Of Drawing Energy From The Sun

First let us ask: Whence comes all the motive power? What is the spring that drives all? We see the ocean rise and fall, the rivers flow, the wind, rain, hail, and snow beat on our windows, the trains and steamers come and go; we here the rattling noise of carriages, the voices from the street; we feel, smell, and taste; and we think of all this. And all this movement, from the surging of the mighty ocean to that subtle movement concerned in our thought, has but one common cause. All this energy emanates from one single centre, one single source, the sun. The sun is the spring that drives all. The sun maintains all human life and supplies all human energy. Another answer we have now found to the above great question: To increase the force accelerating human movement means to turn to the uses of man more of the sun's energy. We honour and revere those great men of bygone times whose names are linked with im-mortal achievements, who have proved themselves benefactors of humanity, the religious reformer with his wise maxims of life, the philosopher with his deep truths, the mathematician with his formulae, the physicist with his laws, the discover with his principles and secrets wrested from nature, the artist with his forms of the beautiful; but who honours him, the greatest of all, who can tell the name of him, who first turned to use the sun's energy to save the effort of a weak fellow creature? That was man's first act of scientific philanthropy, and its consequences have been incalculable.

From the very beginning three ways of drawing energy from the sun were open to man. The savage, when he warmed his frozen limbs at a fire kindled in some way, availed himself of the energy of the sun stored in the burning material. When he carried a bundle of branches to his cave and burned them there, he made use of the sun's stored energy transported from one to another locality. When he set sail to his canoe, he utilized the energy of the sun applied to the atmosphere or the ambient medium. There can be no doubt that the first is the oldest way. A fire, found accidentally, taught the savage to appreciate its beneficial heat. He then very likely conceived of the idea of carrying the glowing members to his abode. Finally, he learned to use the force of a swift current of water or air. It is characteristic of modern development that progress has been effected in the same order. The utilization of the energy stored in wood or coal, or, generally speaking, fuel, led to the steam-engine. Next a great stride in advance was made in energy transportation by the use of electricity, which permitted the transfer of energy from one locality to another without transporting the material. But as to the utilization of the energy of the ambient medium, no radical step forward has as yet been made known.

The ultimate results of development in these three directions are: first, the burning of coal by a cold process in a battery; second, the efficient utilization of the energy of the ambient medium and third the transmission without wires of electrical energy to any distance. In whatever way these results may be arrived at, their practical application will necessarily involve an extensive use of iron, and this invaluable metal will undoubtedly be an essential element in the further development along these three lines. If we succeed in burning coal by a cold process and thus obtain electrical energy in an efficient and inexpensive manner, we shall require in many practical uses of this energy electric motors that is, iron. If we are successful in deriving energy from the ambient medium, we shall need, both in the obtainment and utilization of the energy, machinery again, iron.

If we realize the transmission of electrical energy without wires on an industrial scale, we shall be compelled to use extensively electric generators once more, iron. Whatever we may do, iron will probably be the chief means of accomplishment in the near future, possibly more so than in the past. How long its reign will last is difficult to tell, for even now aluminium is looming up as a threatening competitor. But for the time being, next to providing new resources of energy, it is of the greatest importance to making improvements in the manufacture and utilization of iron. Great advances are possible in these latter directions, which, if brought about, would enormously increase the useful performance of mankind.

# Great Possibilities Offered By Iron For Increasing Human Performance – Enormous Waste In Iron Manufacture

Iron is by far the most important factor in modern progress. It contributes more than any other industrial product to the force accelerating human movement. So general is the use of this metal, and so intimately is it connected with all that concerns our life, that it has become as indispensable to us as the very air we breathe. Its name is synonymous with usefulness. But, however great the influence of iron may be on the present human development, it does not add to the force urging man onward nearly as much as it might. First of all, its manufacture as now carried on is connected with an appalling waste of fuel that is, waste of energy. Then, again, only a part of all the iron produced is applied for useful purposes. A good part of it goes to create frictional resistances, while still another large part is the means of developing negative forces greatly retarding human movement. Thus, the negative force of war is almost wholly represented in iron. It is impossible to estimate with any degree of accuracy the magnitude of this greatest of all retarding forces, but it is certainly very considerable. If the present positive impelling force due to all useful applications of iron be represented by ten, for instance, I should not think it exaggeration to estimate the negative force of war, with due consideration of all its retarding influences and results, at, say, six. On the basis of this estimate the effective impelling force of iron in the positive direction would be measured by the difference of these two numbers, which is four. But if, through the establishment of universal peace, the manufacture of war machinery should cease, and all struggle for supremacy between nations should be turned into healthful, ever active, and productive commercial competition, then the positive impelling force due to iron would be measured by the sum of those two, numbers, which is sixteen that is, this force would have four times its present value. This example is, of course, merely intended to give an idea of the immense increase in the useful performance of mankind which would result from a radical reform of the iron industries supplying the implements of warfare.

A similar inestimable advantage in the saving of energy available to man would be secured by obviating the great waste of coal which is inseparably connected with the present methods of manufacturing iron. In some countries, such as Great Britain, the hurtful effects of this squandering of fuel are beginning to be felt. The price of coal is constantly rising, and the poor are made to suffer more and more. Though we are still far from the dreaded "exhaustion of the coal fields," philanthropy commands us to invent novel methods of manufacturing iron, which will not involve such barbarous waste of this valuable material from which we derive at present most of our energy. It is our duty to coming generations to leave this store of energy intact for them, or at least not to touch it until we shall have perfected processes for burning coal more efficiently. Those who are coming after us will need fuel more than we do. We should be able to manufacture the iron we require by using the sun's energy, without wasting any coal at all. As an effort to this end the idea of smelting iron ores by electric currents obtained from the energy of falling water has naturally suggested itself to many. I have myself spent much time in endeavouring to evolve such a practical process, which would enable iron to be manufactured at small cost. After a prolonged investigation of the subject, finding that it was unprofitable to use the currents generated directly for smelting the ore, I devised a method which is far more economical.

# Economical Production Of Iron By A New Process

The industrial project, as I worked it out six years ago, contemplated the employment of the electric currents derived from the energy of a waterfall, not directly for smelting the ore, but for decomposing water for a preliminary step. To lessen the cost of the plant, I proposed to generate the currents in exceptionally cheap and simple dynamos, which I designed for this sole purpose. The hydrogen liberated in the electrolytic decomposition was to be burned or recombined with oxygen, not with that from which it was separated, but with that of the atmosphere. Thus, very nearly the total electrical energy used up in the decomposition of the water would be recovered in the form of heat resulting from the recombination of the hydrogen. This heat was to be applied to the smelting of ore. The oxygen gained as a by-product of the decomposition of the water I intended to use for certain other industrial purposes, which would probably yield good financial returns, inasmuch as this is the cheapest way of obtaining this gas in large quantities. In any event, it could be employed to burn all kinds of refuse, cheap hydrocarbon, or coal of the most inferior quality which could not be burned in air or be otherwise utilized to advantage, and thus again a considerable amount of heat would be made available for the smelting of the ore. To increase the economy of the process I contemplated, furthermore, using an arrangement such that the hot metal and the products of combustion, coming out of the furnace, would give up their heat upon the cold ore going into the furnace, so that comparatively little of the heat energy would be lost in the smelting. I calculated that probably forty thousand pounds of iron could be produced per horsepower per annum by this method. Liberal allowances were made for those losses which are unavoidable, the above quantity being about half of that theoretically obtainable. Relying on this estimate and on practical data with reference to a certain kind of sand ore existing in abundance in the region of the Great Lakes, including cost of transportation and labour, I found that in some localities iron could be manufactured in this manner cheaper than by any of the adopted methods. This result would be obtained all the more surely if the oxygen obtained from the water, instead of being used for smelting of ore, as assumed, should be more profitably employed. Any new demand for this gas would secure a higher revenue from the plant, thus cheapening the iron. This project was advanced merely in the interest of industry. Someday I hope, a beautiful industrial butterfly will come out of the dusty and shrivelled chrysalis.

The production of iron from sand ores by a process of magnetic separation is highly com-mendable in principle since it involves no waste of coal; but the usefulness of this method is largely reduced by the necessity of melting the iron afterward. As to the crushing of iron ore, I would consider it rational only if done by waterpower, or by energy otherwise obtained without consumption of fuel. An electrolytic cold process, which would make it possible to extract iron cheaply, and also to mold it into the required forms without any fuel consumption, would, in my opinion, be a very great advance in iron manufacture. In common with some other metals, iron has so far resisted electrolytic treatment, but there can be no doubt that such a cold process will ultimately replace in metallurgy the present crude method of casting, and thus obviating the enormous waste of fuel necessitated by the repeated heating of metal in the foundries.

Up to a few decades ago the usefulness of iron was based almost wholly on its remarkable mechanical properties, but since the advent of the commercial dynamo and electric motor its value to mankind has been greatly increased by its unique magnetic qualities. As regards the latter, iron has been greatly improved of late. The signal progress began about thirteen years ago, when I discovered that in using soft Bessemer steel instead of wrought iron, as then customary, in an alternating motor, the performance of the machine was doubled. I brought this fact to the attention of Mr. Albert Schmid, to whose untiring efforts and ability is largely due the supremacy of American electrical machinery, and who was then superintendent of an industrial corporation engaged in this field. Following my suggestion, he constructed transformers of steel, and they showed the same marked improvement. The investigation was then systematically continued under Mr. Schmid's guidance, the impurities being gradually eliminated from the "steel" (which was only such in name, for in reality it was pure soft iron), and soon a product resulted which admitted of little further improvement.

# The Coming Of Age Of Aluminium – Doom Of The Copper Industry – The Great Civilizing Potency Of The New Metal

With the advances made in iron of late years we have arrived virtually at the limits of improvement. We cannot hope to increase very materially its tensile strength, elasticity, hardness, or malleability, nor can we expect to make it much better as regards its magnetic qualities. More recently a notable gain was secured by the mixture of a small percentage of nickel with the iron, but there is not much room for further advance in this direction. New discoveries may be expected, but they cannot greatly add to the valuable properties of the metal, though they may considerably reduce the cost of manufacture. The immediate future of iron is assured by its cheapness and its unrivalled mechanical and magnetic qualities. These are such that no other product can compete with it now. But there can be no doubt that, at a time not very distant, iron, in many of its now uncontested domains, will have to pass the sceptre to another: the coming age will be the age of aluminium. It is only seventy years since this wonderful metal was discovered by Woehler, and the aluminium industry, scarcely forty years old, commands already the attention of the entire world. Such rapid growth has not been recorded in the history of civilization before. Not long-ago aluminium was sold at the fanciful price of thirty or forty dollars per pound; to-day it can be had in any desired amount for as many cents. What is more, the time is not far off when this price, too, will be considered fanciful, for great improvements are possible in the methods of its manufacture. Most of the metal is now produced in the electric furnace by a process combining fusion and electrolysis, which offers a number of advantageous features, but involves naturally a great waste of the electrical energy of the current. My estimates show that the price of aluminium could be considerably reduced by adopting in its manufacture a method similar to that proposed by me for the production of iron. A pound of aluminium requires for fusion only about seventy per cent. of the heat needed for melting a pound of iron, and inasmuch as its weight is only about one third of that of the latter, a volume of aluminium four times that of iron could be obtained from a given amount of heat-energy. But a cold electro-lytic process of manufacture is the ideal solution, and on this I have placed my hope.

The absolutely unavoidable consequence of the advancement of the aluminium industry will be the annihilation of the copper industry. They cannot exist and prosper together, and the latter is doomed beyond any hope of recovery. Even now it is cheaper to convey an electric current through aluminium wires than through copper wires; aluminium castings cost less, and in many domestic and other uses copper has no chance of successfully competing. A further material reduction of the price of aluminium cannot but be fatal to copper. But the progress of the former will not go on unchecked, for, as it ever happens in such cases, the larger industry will absorb the smaller one: the giant copper interests will control the pygmy aluminium interests, and the slow-pacing copper will reduce the lively gait of aluminium. This will only delay, not avoid the impending catastrophe.

Aluminium, however, will not stop at downing copper. Before many years have passed it will be engaged in a fierce struggle with iron, and in the latter, it will find an adversary not easy to conquer. The issue of the contest will largely depend on whether iron shall be indispensable in electric machinery. This the future alone can decide. The magnetism as exhibited in iron is an isolated phenomenon in nature. What it is that makes this metal behave so radically different from all other materials in this respect has not yet been ascertained, though many theories have been suggested. As regards magnetism, the molecules of the various bodies behave like hollow beams partly filled with a heavy fluid and balanced in the middle in the manner of a seesaw. Evidently some disturbing influence exists in nature which causes each molecule, like such a beam, to tilt either one or the other way. If the molecules are tilted one way, the body is magnetic; if they are tilted the other way, the body is non-magnetic; but both positions are stable, as they would be in the case of the hollow beam, owing to the rush of the fluid to the lower end. Now, the wonderful thing is that the molecules of all known bodies went one way, while those of iron went the other way. This metal, it would seem, has an origin entirely different from that of the rest of the globe. It is highly improbable that we shall discover some other and cheaper material which will equal or surpass iron in magnetic qualities.

Unless we should make a radical departure in the character of the electric currents employed, iron will be indispensable. Yet the advantages it offers are only apparent. So long as we use feeble magnetic forces it is by far superior to any other material; but if we find ways of producing great magnetic forces, than better results will be obtain-able without it. In fact, I have already produced electric transformers in which no iron is employed, and which are capable of performing ten times as much work per pound of weight as those of iron. This result is attained by using electric currents of a very high rate of vibration, produced in novel ways, instead of the ordinary currents now employed in the industries. I have also succeeded in operating electric motors without iron by such rapidly vibrating currents, but the results, so far, have been inferior to those obtained with ordinary motors constructed of iron, although theoretically the former should be capable of performing in-comparably more work per unit of weight than the latter. But the seemingly insuperable difficulties which are now in the way may be overcome in the end, and then iron will be done away with, and all electric machinery will be manufactured of aluminium, in all probability, at prices ridiculously low. This would be a severe, if not fatal, blow to iron. In many other branches of industry, as shipbuilding, or wherever lightness of structure is required, the progress of the new metal will be much quicker. For such uses it is eminently suitable and is sure to supersede iron sooner or later. It is highly probable that in the course of time we shall be able to give it many of those qualities which make iron so valuable.

While it is impossible to tell when this industrial revolution will be consummated, there can be no doubt that the future belongs to aluminium, and that in times to come it will be the chief means of increasing human performance. It has in this respect capacities greater by far than those of any other metal. I should estimate its civilizing potency at fully one hundred times that of iron. This estimate, though it may astonish, is not at all exaggerated. First of all, we must remember that there is thirty times as much aluminium as iron in bulk, available for the uses of man. This in itself offers great possibilities. Then, again, the new metal is much more easily workable, which adds to its value. In many of its properties it partakes of the character of a precious metal, which gives it additional worth. Its electric conductivity, which, for a given weight, is greater than that of any other metal, would be alone sufficient to make it one of the most important factors in future human progress. Its extreme lightness makes it far more easy to transport the objects manufactured. By virtue of this property, it will revolutionize naval construction, and in facilitating transport and travel it will add enormously to the useful performance of mankind. But its greatest civilizing property will be, I believe, in aerial travel, which is sure to be brought about by means of it. Telegraphic instruments will slowly enlighten the barbarian. Electric motors and lamps will do it more quickly, but quicker than anything else the flying-machine will do it. By rendering travel ideally easy it will be the best means for unifying the heterogeneous elements of humanity. As the first step toward this realization, we should produce a lighter storage-battery or get more energy from coal.

# Efforts Toward Obtaining More Energy From Coal – The Electric Transmission – The Gas-Engine – The Cold-Coal Battery

I remember that at one time I considered the production of electricity by burning coal in a battery as the greatest achievement toward the advancing civilization, and I am surprised to find how much the continuous study of these subjects has modified my views. It now seems to me that to burn coal, however efficiently, in a battery would be a mere makeshift, a phase in the evolution toward something much more perfect. After all, in generating electricity in this manner, we should be destroying material, and this would be a barbarous process. We ought to be able to obtain the energy we need without consumption of material. But I am far from underrating the value of such an efficient method of burning fuel. At the present time most, motive power comes from coal, and, either directly or by its products, it adds vastly to human energy. Unfortunately, in all the process now adopted, the larger portion of the energy of the coal is uselessly dissipated. The best steam-engines utilize only a small part of the total energy. Even in gas-engines, in which, particularly of late, better results are obtainable, there is still a barbarous waste going on. In our electric-lighting systems we scarcely utilize one third of one per cent., and in lighting by gas a much smaller fraction, of the total energy of the coal. Considering the various uses of coal throughout the world, we certainly do not utilize more than two per cent. of its energy theoretically available. The man who should stop this senseless waste would be a great benefactor of humanity, though the solution he would offer could not be a permanent one, since it would ultimately lead to the exhaustion of the store of material. Efforts toward obtaining more energy from coal are now being made chiefly in two directions by generating electricity and by producing gas for motive-power purposes. In both of these lines notable success has already been achieved.

The advent of the alternating-current system of electric power-transmission marks an epoch in the economy of energy available to man from coal. Evidently all electrical energy obtained from a waterfall, saving so much fuel, is a net gain to mankind, which is all the more effective as it is secured with little expenditure of human effort, and as this most perfect of all known methods of de-riving energy from the sun contributes in many ways to the advancement of civilization. But electricity enables us also to get from coal much more energy than was practicable in the old ways. Instead of transporting the coal to distant places of consumption, we burn it near the mine, develop electricity in the dynamos, and transmit the current to remote localities, thus effecting a considerable saving. Instead of driving the machinery in a factory in the old wasteful way of belts and shafting, we generate electricity by steam-power and operate electric motors. In this manner it is not uncommon to obtain two or three times as much effective motive power from the fuel, besides securing many other important advantages. It is in this field as much as in the transmission of energy to great distance that the alternating system, with its ideally simple machinery, is bringing about an industrial revolution. But in many lines this progress has not been yet fully felt. For example, steamers and trains are still being propelled by the direct application of steam-power to shafts or axles. A much greater percentage of the heat-energy of the fuel could be transformed into motive energy by using, in place of the adopted marine engines and locomotives, dynamos driven by specially designed high-pressure steam- or gas-engines and by utilizing the electricity generated for the propulsion. A gain of fifty to one hundred per cent. in the effective energy derived from the coal could be secured in this manner. It is difficult to understand why a fact so plain and obvious is not receiving more attention from engineers. In ocean steamers such an improvement would be particularly desirable, as it would do away with noise and increase materially the speed and the carrying capacity of the liners.

Still more energy is now being obtained from coal by the latest improved gas-engine, the economy of which is, on the average, probably twice that of the best steam-engine. The introduction of the gas-engine is very much facilitated by the importance of the gas industry. With the increasing use of the electric light more and more of the gas is utilized for heating and motive-power purposes. In many instances gas is manufactured close to the coal-mine and conveyed to distant places of consumption, a considerable saving both in cost of transportation and in utilization of the energy of the fuel being thus effected. In the present state of the mechanical and electrical arts the most rational way of deriving energy from coal is evidently to manufacture gas close to the coal store, and to utilize it, either on the spot or elsewhere, to generate electricity for industrial uses in dynamos driven by gas engines. The commercial success of such a plant is largely dependent upon the production of gas-engines of great nominal horsepower, which, judging from the keen activity in this field will soon be forthcoming. Instead of consuming coal directly, as usual, gas should be manufactured from it and burned to economize energy.

But all such improvements cannot be more than passing phases in the evolution toward some-thing far more perfect, for ultimately, we must succeed in obtaining electricity from coal in a more direct way, involving no great loss of heat-energy. Whether coal can be oxidized by a cold process is still a question. Its combination with oxygen always involves heat, and whether the energy of the combination of the carbon with another element can be turned directly into electrical energy has not yet been determined. Under certain conditions nitric acid will burn the carbon, generating an electric current, but the solution does not remain cold. Other means of oxidizing coal have been proposed, but they have offered no promise of leading to an efficient process. My own lack of success has been complete, though perhaps not quite so complete as that of some who have "perfected" the cold-coal battery. This problem is essentially one for the chemist to solve. It is not for the physicist, who determines all his results in advance, so that, when the experiment is tried, it cannot fail. Chemistry, though a positive science, does not yet admit of a solution by such positive methods as those which are available in the treatment of many physical problems. The result, if possible, will be arrived at through patent trying rather than through deduction or calculation. The time will soon come, however, when the chemist will be able to follow a course clearly mapped out beforehand, and when the process of his arriving at a desired result will be purely constructive. The cold-coal battery would give a great impetus to electrical development; it would lead very shortly to a practical flying-machine and would enormously enhance the introduction of the automobile. But these and many other problems will be better solved, and in a more scientific manner, by a light storage battery.

# Energy From The Medium – The Windmill And The Solar Engine – Motive Power From Terrestrial Heat – Electricity From Natural Sources

Besides fuel, there is abundant material from which we might eventually derive power. An immense amount of energy is locked up in limestone, for instance, and machines can be driven by liberating the carbonic acid through sulphuric acid or otherwise. I once constructed such an engine, and it operated satisfactorily.

But whatever our resources of primary energy may be in the future, we must, to be rational, obtain it without consumption of any material. Long ago I came to this conclusion, and to arrive at this result only two ways, as before indicated, appeared possible either to turn to use the energy of the sun stored in the ambient medium, or to transmit, through the medium, the sun's energy to distant places from some locality where it was obtainable without consumption of material. At that time, I at once rejected the latter method as entirely impracticable and turned to examine the possibilities of the former.

It is difficult to believe, but it is, nevertheless, a fact, that since time immemorial man has had at his disposal a fairly good machine which has enabled him to utilize the energy of the ambient medium. This machine is the windmill. Contrary to popular belief, the power obtainable from wind is very considerable. Many a deluded inventor has spent years of his life in endeavouring to "harness the tides," and some have even proposed to com-press air by tide- or wave-power for supplying energy, never understanding the signs of the old windmill on the hill, as it sorrowfully waved its arms about and bade them stop. The fact is that a wave- or tide-motor would have, as a rule, but a small chance of competing commercially with the windmill, which is by far the better machine, allowing a much greater amount of energy to be obtained in a simpler way. Wind-power has been, in old times, of inestimable value to man, if for nothing else but for enabling him, to cross the seas, and it is even now a very important factor in travel and transportation. But there are great limitations in this ideally simple method of utilizing the sun's energy. The machines are large for a given output, and the power is intermittent, thus necessitating the storage of energy and increasing the cost of the plant.

A far better way, however, to obtain power would be to avail ourselves of the sun's rays, which beat the earth incessantly and supply energy at a maximum rate of over four million horsepower per square mile. Although the average energy received per square mile in any locality during the year is only a small fraction of that amount, yet an inexhaustible source of power would be opened up by the discovery of some efficient method of utilizing the energy of the rays. The only rational way known to me at the time when I began the study of this subject was to employ some kind of heat- or thermodynamic-engine, driven by a volatile fluid evaporate in a boiler by the heat of the rays. But closer investigation of this method, and calculation, showed that, notwithstanding the apparently vast amount of energy received from the sun's rays, only a small fraction of that energy could be actually utilized in this manner. Furthermore, the energy supplied through the sun's radiations is periodical, and the same limitations as in the use of the windmill I found to exist here also. After a long study of this mode of obtaining motive power from the sun, taking into account the necessarily large bulk of the boiler, the low efficiency of the heat-engine, the additional cost of storing the energy and other drawbacks, I came to the conclusion that the "solar engine," a few instances excepted, could not be industrially exploited with success.

Another way of getting motive power from the medium without consuming any material would be to utilize the heat contained in the earth, the water, or the air for driving an engine. It is a well-known fact that the interior portions of the globe are very hot, the temperature rising, as observations show, with the approach to the centre at the rate of approximately 1 degree C. for every hundred feet of depth. The difficulties of sinking shafts and placing boilers at depths of, say, twelve thousand feet, corresponding to an increase in temperature of about 120 degrees C., are not insuperable, and we could certainly avail ourselves in this way of the internal heat of the globe. In fact, it would not be necessary to go to any depth at all in order to derive energy from the stored terrestrial heat. The superficial layers of the earth and the air strata close to the same are at a temperature sufficiently high to evaporate some extremely volatile substances, which we might use in our boilers instead of water. There is no doubt that a vessel might be propelled on the ocean by an engine driven by such a volatile fluid, no other energy being used but the heat abstracted from the water. But the amount of power which could be obtained in this manner would be, without further provision, very small.

Electricity produced by natural causes is another source of energy which might be rendered available. Lightning discharges involve great amounts of electrical energy, which we could utilize by transforming and storing it. Some years ago, I made known a method of electrical transformation which renders the first part of this task easy, but the storing of the energy of lightning discharges will be difficult to accomplish. It is well known, furthermore, that electric currents circulate constantly through the earth, and that there exists between the earth and any air stratum a difference of electrical pressure, which varies in proportion to the height.

In recent experiments I have discovered two novel facts of importance in this connection. One of these facts is that an electric current is generated in a wire extending from the ground to a great height by the axial, and probably also by the translatory, movement of the earth. No appreciable current, however, will flow continuously in the wire unless the electricity is allowed to leak out into the air. Its escape is greatly facilitated by providing at the elevated end of the wire a conducting terminal of great surface, with many sharp edges or points. We are thus enabled to get a continuous supply of electrical energy by merely supporting a wire at a height, but, unfortunately, the amount of electricity which can be so obtained is small.

The second fact which I have ascertained is that the upper air strata are permanently charged with electricity opposite to that of the earth. So, at least, I have interpreted my observations, from which it appears that the earth, with its adjacent insulating and outer conducting envelope, constitutes a highly charged electrical condenser containing, in all probability, a great amount of electrical energy which might be turned to the uses of man, if it were possible to reach with a wire to great altitudes.

It is possible, and even probable, that there will be, in time, other resources of energy opened up, of which we have no knowledge now. We may even find ways of applying forces such as magnetism or gravity for driving machinery without using any other means. Such realizations, though highly improbable, are not impossible. An example will best convey an idea of what we can hope to attain and what we can never attain. Imagine a disk of some homogeneous material turned perfectly true and arranged to turn in frictionless bearings on a horizontal shaft above the ground. This disk, being under the above conditions perfectly balanced, would rest in any position. Now, it is possible that we may learn how to make such a disk rotate continuously and perform work by the force of gravity without any further effort on our part; but it is perfectly impossible for the disk to turn and to do work without any force from the outside. If it could do so, it would be what is designated scientifically as a "perpetuum mobile," a machine creating its own motive power. To make the disk rotate by the force of gravity we have only to invent a screen against this force. By such a screen we could prevent this force from acting on one half of the disk, and the rotation of the latter would follow. At least, we cannot deny such a possibility until we know exactly the nature of the force of gravity. Suppose that this force were due to a movement comparable to that of a stream of air passing from above toward the centre of the earth. The effect of such a stream upon both halves of the disk would be equal, and the latter would not rotate ordinarily; but if one half should be guarded by a plate arresting the movement, then it would turn.

# A Departure From Known Methods – Possibility Of A "Self-Acting" Engine Or Machine, Inanimate, Yet Capable, Like A Living Being, Of Deriving En-Ergy From The Medium – The Ideal Way Of Obtaining Motive Power

When I began the investigation of the subject under consideration, and when the preceding or similar ideas presented themselves to me for the first time, though I was then unacquainted with a number of the facts mentioned, a survey of the various ways of utilizing the energy of the medium convinced me, nevertheless, that to arrive at a thoroughly satisfactory practical solution a radical departure from the methods then known had to be made. The windmill, the solar engine, the engine driven by terrestrial heat, had their limitations in the amount of power obtainable. Some new way had to be discovered which would enable us to get more energy. There was enough heat-energy in the medium, but only a small part of it was available for the operation of an engine in the ways then known. Besides, the energy was obtainable only at a very slow rate. Clearly, then, the problem was to discover some new method which would make it possible both to utilize more of the heat-energy of the medium and also to draw it away from the same at a more rapid rate.

I was vainly endeavouring to form an idea of how this might be accomplished, when I read some statements from Carnot and Lord Kelvin (then Sir William Thomson) which meant virtually that it is impossible for an inanimate mechanism or self-acting machine to cool a portion of the medium below the temperature of the surrounding and operate by the heat abstracted. These statements interested me intensely. Evidently a living being could do this very thing, and since the experiences of my early life which I have related had convinced me that a living being is only an automaton, or, otherwise stated, a "self-acting-engine," I came to the conclusion that it was possible to construct a machine which would do the same. As the first step toward this realization, I conceived the following mechanism. Imagine a thermopile consisting of a number of bars of metal extending from the earth to the outer space beyond the atmosphere. The heat from below, conducted upward along these metal bars, would cool the earth or the sea or the air, according to the location of the lower parts of the bars, and the result, as is well known, would be an electric current circulating in these bars. The two terminals of the thermopile could now be joined through an electric motor, and, theoretically, this motor would run on and on, until the media below would be cooled down to the temperature of the outer space. This would be an inanimate engine which, to all evidence, would be cooling a portion of the medium below the temperature of the surrounding, and operating by the heat abstracted.

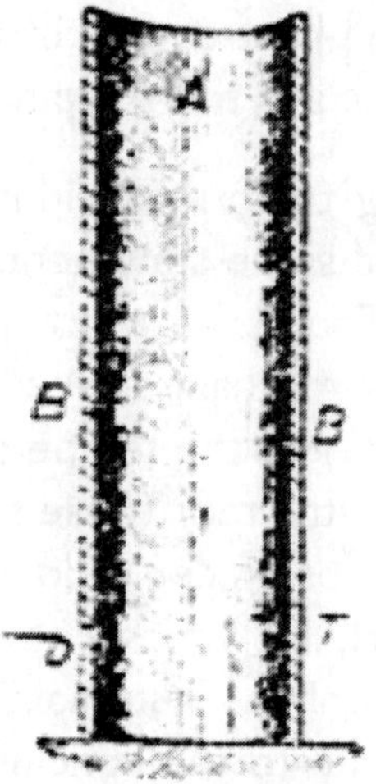


**Diagram b:Obtaining energy from the ambient medium A, medium with little energy; B, B, ambient medium with much energy; O, path of the energy.**

But was it not possible to realize a similar condition without necessarily going to a height? Conceive, for the sake of illustration, (a cylindrical) enclosure T, as illustrated in diagram b, such that energy could not be transferred across it except through a channel or path O, and that, by some means or other, in this enclosure a medium were maintained which would have little energy, and that on the outer side of the same there would be the ordinary ambient medium with much energy. Under these assumptions the energy would flow through the path O, as indicated by the arrow, and might then be converted on its passage into some other form of energy. The question was, Could such a condition be attained? Could we produce artificially such a "sink" for the energy of the ambient medium to flow in? Suppose that an extremely low temperature could be maintained by some process in a given space; the surrounding medium would then be compelled to give off heat, which could be converted into mechanical or other form of energy and utilized. By realizing such a plan, we should be enabled to get at any point of the globe a continuous supply of energy, day, and night. More than this, reasoning in the abstract, it would seem possible to cause a quick circulation of the medium, and thus draw the energy at a very rapid rate.

Here, then, was an idea which, if realizable, afforded a happy solution of the problem of getting energy from the medium. But was it realizable? I convinced myself that it was so in a number of ways, of which one is the following. As regards heat, we are at a high level, which may be represented by the surface of a mountain lake considerably above the sea, the level of which may mark the absolute zero of temperature existing in the interstellar space. Heat, like water, flows from high to low level, and, consequently, just as we can let the water of the lake run down to the sea, so we are able to let heat from the earth's surface travel up into the cold region above. Heat, like water, can perform work in flowing down, and if we had any doubt as to whether we could derive energy from the medium by means of a thermopile, as before described, it would be dispelled by this analogue. But can we produce cold in a given portion of the space and cause the heat to flow in continually? To create such a "sink," or "cold hole," as we might say, in the medium, would be equivalent to producing in the lake a space either empty or filled with something much lighter than water. This we could do by placing in the lake a tank and pumping all the water out of the latter. We know, then, that the water, if allowed to flow back into the tank, would, theoretically, be able to perform exactly the same amount of work which was used in pumping it out, but not a bit more. Consequently, nothing could be gained in this double operation of first raising the water and then letting it fall down. This would mean that it is impossible to create such a sink in the medium. But let us reflect a moment. Heat, though following certain general laws of mechanics, like a fluid, is not such; it is energy which may be converted into other forms of energy as it passes from a high to a low level. To make our mechanical analogy complete and true, we must, therefore, assume that the water, in its passage into the tank, is converted into something else, which may be taken out of it without using any, or by using very little, power. For example, if heat be represented in this analogue by the water of the lake, the oxygen and hydrogen composing the water may illustrate other forms of energy into which the heat is transformed in passing from hot to cold. If the process of heat transformation were absolutely perfect, no heat at all would arrive at the low level since all of it would be converted into other forms of energy. Corresponding to this ideal case, all the water flowing into the tank would be decomposed into oxygen and hydrogen before reaching the bottom, and the result would be that water would continually flow in, and yet the tank would remain entirely empty, the gases formed escaping. We would thus produce, by expending initially a certain amount of work to create a sink for the heat or, respectively, the water to flow in, a condition enabling us to get any amount of energy without further effort. This would be an ideal way of obtaining motive power. We do not know of any such absolutely perfect process of heat-conversion, and consequently some heat will generally reach the low level, which means to say, in our mechanical analogue, that some water will arrive at the bottom of the tank, and a gradual and slow filling of the latter will take place, necessitating continuous pumping out. But evidently there will be less to pump out than flows in, or, in other words, less energy will be needed to maintain the initial condition than is developed by the fall, and this is to say that some energy will be gained from the medium. What is not converted in flowing down can just be raised up with its own energy, and what is converted is clear gain. Thus, the virtue of the principle I have discovered resides wholly in the conversion of the energy on the downward flow.

# First Efforts To Produce The Self-Acting Engine – The Mechanical Oscillator – Work Of Dewar And Linde Liquid Air

Having recognized this truth, I began to devise means for carrying out my idea, and, after long thought, I finally conceived a combination of apparatus which should make possible the obtaining of power from the medium by a process of continuous cooling of atmospheric air. This apparatus, by continually transforming heat into mechanical work, tended to become colder and colder, and if it only were practicable to reach a very low temperature in this manner, then a sink for the heat could be produced, and energy could be derived from the medium. This seemed to be contrary to the statements of Carnot and Lord Kelvin before referred to, but I concluded from the theory of the process that such a result could be attained. This conclusion I reached, I think, in the latter part of 1883, when I was in Paris, and it was at a time when my mind was being more and more dominated by an invention which I had evolved during the preceding year, and which has since become known under the name of the "rotating magnetic field." During the few years which followed I elaborated further the plan I had imagined, and studied the working conditions, but made little headway. The commercial introduction in this country of the invention before referred to required most of my energies until 1889, when I again took up the idea of the self-acting machine. A closer investigation of the principals involved, and calculation, now showed that the result I aimed at could not be reached in a practical manner by ordinary machinery, as I had in the beginning expected. This led me, as a next step, to the study of a type of engine generally designated as "tur-bine," which at first seemed to offer better chances for a realization of the idea. Soon I found, how- ever, that the turbine, too, was unsuitable. But my conclusions showed that if an engine of a peculiar kind could be brought to a high degree of perfection, the plan I had conceived was realizable, and I resolved to proceed with the development of such an engine, the primary object of which was to secure the greatest economy of transformation of heat into mechanical energy. A characteristic feature of the engine was that the work-performing piston was not connected with anything else but was perfectly free to vibrate at an enormous rate. The mechanical difficulties encountered in the construction of this engine were greater than I had anticipated, and I made slow progress. This work was continued until early in 1892, when I went to London, where I saw Professor Dewar's admirable experiments with liquefied gases. Others had liquefied gases before, and notably Ozlewski and Pictet had performed creditable early experiments in this line, but there was such a vigour about the work of Dewar that even the old appeared new. His experiments showed, though in a way different from that I had imagined, that it was possible to reach a very low temperature by transforming heat into mechanical work, and I returned, deeply im-pressed with what I had seen, and more than ever convinced that my plan was practicable. The work temporarily interrupted was taken up anew, and soon I had in a fair state of perfection the engine which I have named "the mechanical oscillator." In this machine I succeeded in doing away with all packings, valves, and lubrication, and in producing so rapid a vibration of the piston that shafts of tough steel, fastened to the same and vibrated longitudinally, were torn asunder. By combining this engine with a dynamo of special design I produced a highly efficient electrical generator, invaluable in measurements and determinations of physical quantities on account of the unvarying rate of oscillation obtainable by its means. I exhibited several types of this machine, named "mechanical and electrical oscillator," before the Electrical Congress at the World's Fair in Chicago during the summer of 1893, in a lecture which, on account of other pressing work, I was unable to prepare for publication. On that occasion I exposed the principles of the mechanical oscillator, but the original purpose of this machine is explained here for the first time.

In the process, as I had primarily conceived it, for the utilization of the energy of the ambient medium, there were five essential elements in combination, and each of these had to be newly designed and perfected, as no such machines existed. The mechanical oscillator was the first element of this combination, and having perfected this, I turned to the next, which was an air-compressor of a design in certain respects resembling that of the mechanical oscillator. Similar difficulties in the construction were again encountered, but the work was pushed vigorously, and at the close of 1894 I had completed these two elements of the combination, and thus produced an apparatus for compressing air, virtually to any desired pressure, incomparably simpler, smaller, and more efficient than the ordinary. I was just beginning work on the third element, which together with the first two would give a refrigerating machine of exceptional efficiency and simplicity, when a misfortune befell me in the burning of my laboratory, which crippled my labours and delayed me. Shortly afterward Dr. Carl Linde announced the liquefaction of air by a self-cooling process, demonstrating that it was practicable to proceed with the cooling until liquefaction of the air took place. This was the only experimental proof which I was still wanting that energy was obtainable from the medium in the manner contemplated by me.

The liquefaction of air by a self-cooling process was not, as popularly believed, an accidental discovery, but a scientific result which could not have been delayed much longer, and which, in all probability, could not have escaped Dewar. This fascinating advance, I believe, is largely due to the powerful work of this great Scotchman. Nevertheless, Linde's is an immortal achievement. The manufacture of liquid air has been carried on for four years in Germany, on a scale much larger than in any other country, and this strange product has been applied for a variety of purposes. Much was expected of it in the beginning, but so far it has been an industrial ignis fatuus. By the use of such machinery as I am perfecting, its cost will probably be greatly lessened, but even then, its commercial success will be questionable. When, used as a refrigerant it is uneconomical, as its temperature is unnecessarily low. It is as expensive to maintain a body at a very low temperature as it is to keep it very hot; it takes coal to keep air cold. In oxygen manufacture it cannot yet compete with the electrolytic method. For use as an explosive, it is unsuitable, because its low temperature again condemns it to a small efficiency, and for motive power purposes its cost is still by far too high. It is of interest to note, however, that in driving an engine by liquid air a certain amount of energy may be gained from the engine, or stated otherwise, from the ambient medium which keeps the engine warm, each two hundred pounds of iron casting of the latter contributing energy at the rate of about one effective horsepower during one hour. But this gain of the consumer is offset by an equal loss of the producer.

Much of this task on which I have laboured so long remains to be done. A number of mechanical details are still to be perfected and some difficulties of a different nature to be mastered, and I cannot hope to produce a self-acting machine deriving energy from the ambient medium for a long time yet, even if all my expectations should materialize. Many circumstances have occurred which have retarded my work of late, but for several reasons the delay was beneficial.

One of these reasons was that I had ample time to consider what the ultimate possibilities of this development might be. I worked for a long time fully convinced that the practical realization of this method of obtaining energy from the sun would be of incalculable industrial value, but the continued study of the subject revealed the fact that while it will be commercially profitable if my expectations are well founded, it will not be so to an extraordinary degree.

# Discovery Of Unexpected Properties Of The Atmosphere – Strange Experiments – Transmission Of Electrical Energy Through One Wire Without Return – Transmission Through The Earth Without Any Wire

Another of these reasons was that I was led to recognize the transmission of electrical energy to any distance through the media as by far the best solution of the great problem of harnessing the sun's energy for the uses of man. For a long time, I was convinced that such a transmission on an industrial scale, could never be realized, but a discovery which I made changed my view. I observed that under certain conditions the atmosphere, which is normally a high insulator, assumes con-ducting properties, and so becomes capable of conveying any amount of electrical energy. But the difficulties in the way of a practical utilization of this discovery for the purpose of transmitting electrical energy without wires were seemingly insuperable. Electrical pressures of many millions of volts had to be produced and handled; generating apparatus of a novel kind, capable of withstanding the immense electrical stresses, had to be invented and perfected, and a complete safety against the dangers of the high-tension currents had to be attained in the system before its practical introduction could be even thought of. All this could not be done in a few weeks or months, or even years. The work required patience and constant application, but the improvements came, though slowly. Other valuable results were, however, arrived at in the course of this long-continued work, of which I shall endeavour to give a brief account, enumerating the chief advances as they were successively effected

The discovery of the conducting properties of the air, though unexpected, was only a natural result of experiments in a special field which I had carried on for some years before. It was, I believe, during 1889 that certain possibilities offered by extremely rapid electrical oscillations determined me to design a number of special machines adapted for their investigation. Owing to the peculiar requirements, the construction of these machines was very difficult, and consumed much time and effort; but my work on them was generously rewarded, for I reached by their means several novel and important results. One of the earliest observations I made with these new machines was that electrical oscillations of an extremely high-rate act in an extraordinary manner upon the human organism. Thus, for instance, I demonstrated that powerful electrical discharges of several hundred thousand volts, which at that time were considered absolutely deadly, could be passed through the body without inconvenience or hurtful consequences. These oscillations produced other specific physiological effects, which, upon my announcement, were eagerly taken up by skilled physicians and further investigated. This new field has proved itself fruitful beyond expectation, and in the few years which have passed since, it has been developed to such an extent that it now forms a legitimate and important department of med Many results, thought impossible at that time, are now readily obtainable with these oscillations, and many experiments undreamed of then can now be readily performed by their means. I still remember with pleasure how nine years ago, I passed the discharge of a powerful induction-coil through my body to demonstrate before a scientific society the comparative harmlessness of very rapidly vibrating electric currents, and I can still recall the astonishment of my audience. I would now undertake, with much less apprehension that I had in that experiment, to transmit through my body with such currents the entire electrical energy of the dynamos now working at Niagara forty or fifty thousand horsepower. I have produced electrical oscillations which were of such intensity that when circulating through my arms and chest they have melted wires which joined my hands, and still I felt no inconvenience. I have energized with such oscillations a loop of heavy copper wire so powerfully that masses of metal, and even objects of an electrical resistance specifically greater than that of human tissue brought close to or placed within the loop, were heated to a high temperature and melted, often with the violence of an explosion, and yet into this very space in which this terribly-destructive turmoil was going on I have repeatedly thrust my head without feeling anything or experiencing injurious after-effects.

Another observation was that by means of such oscillations light could be produced in a novel and more economical manner, which promised to lead to an ideal system of electric illumination by vacuum-tubes, dispensing with the necessity of renewal of lamps or incandescent filaments, and possibly also with the use of wires in the interior of buildings. The efficiency of this light increases in proportion to the rate of the oscillations, and its commercial success is, therefore, dependent on the economical production of electrical vibrations of transcending rates. In this direction I have met with gratifying success of late, and the practical introduction of this new system of illumination is not far off.

The investigations led to many other valuable observations and results, one of the more important of which was the demonstration of the practicability of supplying electrical energy through one wire without return. At first, I was able to transmit in this novel manner only very small amounts of electrical energy, but in this line also my efforts have been rewarded with similar success.

An ordinary incandescent lamp, connected with one or both of its terminals to the wire forming the upper free end of the coil shown in the photograph, is lighted by electrical vibrations conveyed to it through the coil from an electrical oscillator, which is worked only to one fifth of one per cent. of its full capacity.

**Fig. 3. Experiment to illustrate the supplying of electrical energy through a single wire without return.**

The photograph shown in Fig. 3 illustrates, as its title explains, an actual transmission of this kind effected with apparatus used in other experiments here described. To what a degree the appliances have been perfected since my first demonstrations early in 1891 before a scientific society, when my apparatus was barely capable of lighting one lamp (which result was considered wonderful), will appear when I state that I have now no difficulty in lighting in this manner four or five hundred lamps, and could light many more. In fact, there is no limit to the amount of energy which may in this way be supplied to operate any kind of electrical device.

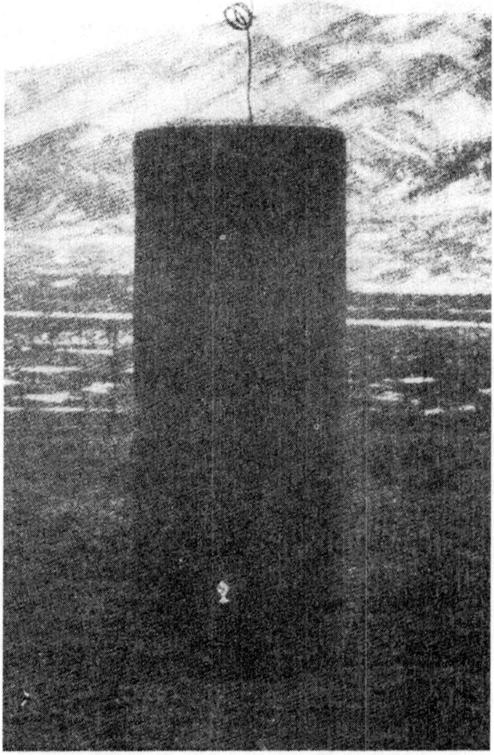

**Fig. 4: Experiment to illustrate the transmission of electrical energy through the Earth without wire.**

The coil shown in the photograph has its lower end or terminal connected to the ground and is exactly attuned to the vibrations of a distant electrical oscillator. The lamp lighted is in an independent wire loop, energized by induction from the coil excited by the electrical vibrations transmitted to it through the ground from the oscillator, which is worked only to five per cent. of its full capacity.

After demonstrating the practicability of this method of transmission, the thought naturally occurred to me to use the earth as a conductor, thus dispensing with all wires. Whatever electricity may be, it is a fact that it behaves like an incompressible fluid, and the earth may be looked upon as an immense reservoir of electricity, which, I thought, could be disturbed effectively by a properly designed electrical machine. Accordingly, my next efforts were directed toward perfecting a special apparatus which would be highly effective in creating a disturbance of electricity in the earth. The progress in this new direction was necessarily very slow and the work discouraging, until I finally succeeded in perfecting a novel kind of transformer or induction-coil, particularly suited for this special purpose. That it is practicable, in this manner, not only to transmit minute amounts of electrical energy for operating delicate electrical devices, as I contemplated at first, but also electrical energy in appreciable quantities, will appear from an inspection of Fig. 4, which illustrates an actual experiment of this kind performed with the same apparatus. The result obtained was all the more remarkable as the top end of the coil was not connected to a wire or plate for magnifying the effect.

# Wireless" Telegraphy – The Secret Of Tuning – Errors In The Hertzian Investigations – A Receiver Of Wonderful Sensitiveness

As the first valuable result of my experiments in this latter line a system of telegraphy without wires resulted, which I described in two scientific lectures in February and March 1893. It is mechanically illustrated in diagram c, the upper part of which shows the electrical arrangement as I described it then, while the lower part illustrates its mechanical analogue. The system is extremely simple in principle. Imagine two tuning-forks F, F1, one at the sending- and the other at the receiving station respectively, each having attached to its lower prong a minute piston p, fitting in a cylinder. Both the cylinders communicate with a large reservoir R, with elastic walls, which is supposed to be closed and filled with a light and incompressible fluid. By striking repeatedly one of the prongs of the tuning-fork F, the small piston p below would be vibrated, and its vibrations, transmitted through the fluid, would reach the distant fork F1, which is "tuned" to the fork F, or stated otherwise, of exactly the same note as the latter. The fork F1 would now be set vibrating, and its vibration would be intensified by the continued action of the distant fork F until its upper prong, swinging far out, would make an electrical connection with a stationary contact c", starting in this manner some electrical or other appliances which may be used for recording the signals. In this simple way messages could be exchanged between the two stations, a similar contact c' being provided for this purpose, close to the upper prong of the fork F, so that the apparatus at each station could be employed in turn as receiver and transmitter.

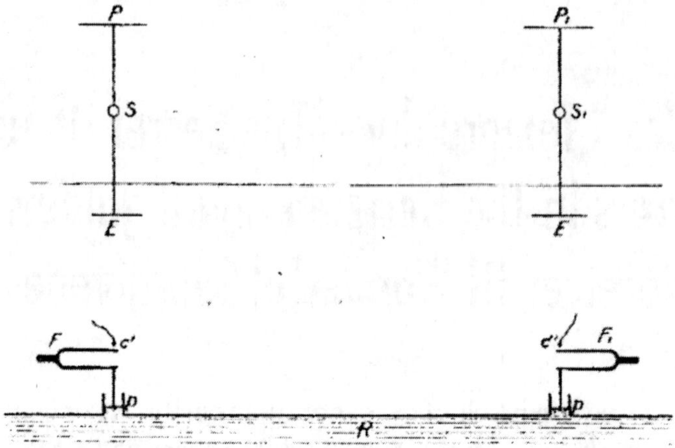


**Diagram c: "Wireless" telegraphy mechanically illustrated.**

The electrical system illustrated in the upper figure of diagram c is exactly the same in principle, the two wires or circuits ESP and E1S1P1, which extend vertically to a height, representing the two tuning-forks with the pistons attached to them. These circuits are connected with the ground by plates E, E1, and to two elevated metal sheets P, P1, which store electricity and thus magnify considerably the effect. The closed reservoir R, with elastic walls, is in this case replaced by the earth, and the fluid by electricity. Both of these circuits are "tuned" and operate just like the two tuning forks. Instead of striking the fork F at the sending station, electrical oscillations are produced in the vertical sending- or transmitting-wire ESP, as by the action of a source S, included in this wire, which spread through the ground and reach the distant vertical receiving-wire E1S1P1, exciting corresponding electrical oscillations in the same. In the latter wire or circuit is included a sensitive device or receiver S1, which is thus set in action and made to operate a relay or other appliance. Each station is, of course, provided both with a source of electrical oscillations S and a sensitive receiver S1, and a simple provision is made for using each of the two wires alternately to send and to receive the messages.

**Fig.5: Photographic view of the coils responding to electrical oscillations.**

The picture shows a number of coils, differently attuned and responding to the vibrations transmitted to them through the earth from an electrical oscillator. The large coil on the right, discharging strongly, is tuned to the fundamental vibration, which is fifty thousand per second; the two larger vertical coils to twice that number; the smaller white wire coil to four times that number, and the remaining small coils to higher tones. The vibrations produced by the oscillator were so intense that they affected perceptibly a small coil tuned to the twenty-sixth higher tone.

The exact attunement of the two circuits secures great advantages, and, in fact, it is essential in the practical use of the system. In this respect many popular errors exist, and, as a rule, in the technical reports on these subject circuits and appliances are described as affording these advantages when from their very nature it is evident that this is impossible. In order to attain the best results, it is essential that the length of each wire or circuit, from the ground connection to the top, should be equal to one quarter of the wavelength of the electrical vibration in the wire, or else equal to that length multiplied by an odd number. Without the observation of this rule, it is virtually impossible to prevent the interference and insure the privacy of messages. Therein lies the secret of tuning. To obtain the most satisfactory results it is, however, necessary to resort to electrical vibrations of low pitch. The Hertzian spark apparatus, used generally by experimenters, which produces oscillations of a very high rate, permits no effective tuning, and slight disturbances are sufficient to render an exchange of messages impracticable. But scientifically designed, efficient appliances allow nearly perfect adjustment. An experiment performed with the improved apparatus repeatedly referred to, and intended to convey an idea of this feature, is illustrated in Fig. 5, which is sufficiently explained by its note.

Since I described these simple principles of telegraphy without wires, I have had frequent occasion to note that the identical features and elements have been used, in the evident belief that the signals are being transmitted to considerable distance by "Hertzian" radiations. This is only one of many misapprehensions to which the investigations of the lamented physicist have given rise. About thirty-three years ago Maxwell, following up a suggestive experiment made by Faraday in 1845, evolved an ideally simple theory which intimately connected light, radiant heat, and electrical phenomena, interpreting them as being all due to vibrations of a hypothetical fluid of inconceivable tenuity, called the ether. No experimental verification was arrived at until Hertz, at the suggestion of Helmholtz, undertook a series of experiments to this effect. Hertz proceeded with extraordinary ingenuity and insight, but devoted little energy to the perfection of his old-fashioned apparatus. The consequence was that he failed to observe the important function which the air played in his experiments, and which I subsequently discovered. Repeating his experiments and reaching different results, I ventured to point out this oversight. The strength of the proofs brought forward by Hertz in support of Maxwell's theory resided in the correct estimate of the rates of vibration of the circuits he used. But I ascertained that he could not have obtained the rates he thought he was getting. The vibrations with identical apparatus he employed are, as a rule, much slower, this being due to the presence of air, which produces a dampening effect upon a rapidly vibrating electric circuit of high pressure, as a fluid does upon a vibrating tuning-fork. I have, however, discovered since that time other causes of error, and I have long ago ceased to look upon his results as being an experimental verification of the poetical conceptions of Maxwell. The work of the great German physicist has acted as an immense stimulus to contemporary electrical research, but it has likewise, in a measure, by its fascination, paralyzed the scientific mind, and thus hampered independent inquiry. Every new phenomenon which was discovered was made to fit the theory, and so very often the truth has been unconsciously distorted.

**Fig. 6: Photographic view of the essential parts of the electrical oscillator used in the experiments described.**

When I advanced this system of telegraphy, my mind was dominated by the idea of effecting communication to any distance through the earth or environing medium, the practical consummation of which I considered of transcendent importance, chiefly on account of the moral effect which it could not fail to produce universally. As the first effort to this end, I proposed at that time, to employ relay-stations with tuned circuits, in the hope of making thus practicable signalling over vast distances, even with apparatus of very moderate power then at my command. I was confident, however, that with properly designed machinery signals could be transmitted to any point of the globe, no matter what the distance, without the necessity of using such intermediate stations. I gained this conviction through the discovery of a singular electrical phenomenon, which I described early in 1892, in lectures I delivered before some scientific societies abroad, and which I have called a "rotating brush." This is a bundle of light which is formed, under certain conditions, in a vacuumbulb, and which is of a sensitiveness to magnetic and electric influences bordering, so to speak, on the supernatural. This light-bundle is rapidly rotated by the earth's magnetism as many as twenty thousand times per second, the rotation in these parts being opposite to what it would be in the southern hemisphere, while in the region of the magnetic equator it should not rotate at all. In its most sensitive state, which is difficult to obtain, it is responsive to electric or magnetic influences on an incredible degree. The mere stiffening of the muscles of the arm and consequent slight electrical change in the body of an observer standing at some distance from it, will perceptibly affect it. When in this highly sensitive state it is capable of indicating the slightest magnetic and electric changes taking place in the earth. The observation of this wonderful phenomenon impressed me strongly that communication at any distance could be easily effected by its means, provided that apparatus could be perfected capable of producing an electric or magnetic change of state, however small, in the terrestrial globe or environing medium.

# Development Of A New Principle – The Electrical Oscillator – Production Of Immense Electrical Movements – The Earth Responds To Man – Interplanetary Communication Now Probable

I resolved to concentrate my efforts upon this venturesome task, though it involved great sacrifice, for the difficulties to be mastered were such that I could hope to consummate it only after years of labour. It meant delay of other work to which I would have preferred to devote myself, but I gained the conviction that my energies could not be more usefully employed; for I recognized that an efficient apparatus for, the production of powerful electrical oscillations, as was needed for that specific purpose, was the key to the solution of other most important electrical and, in fact, human problems. Not only was communication, to any distance, without wires possible by its means, but, likewise, the transmission of energy in great amounts, the burning of the atmospheric nitrogen, the production of an efficient illuminate, and many other results of inestimable scientific and industrial value. Finally, however, I had the satisfaction of accomplishing the task undertaken by the use of a new principle, the virtue of which is based on the marvellous properties of the electrical condenser. One of these is that it can discharge or explode its stored energy in an inconceivably short time. Owing to this it is unequalled in explosive violence. The explosion of dynamite is only the breath of a consumptive compared with its discharge. It is the means of producing the strongest current, the highest electrical pressure, the greatest commotion in the medium. Another of its properties, equally valuable, is that its discharge may vibrate at any rate desired up to many millions per second.

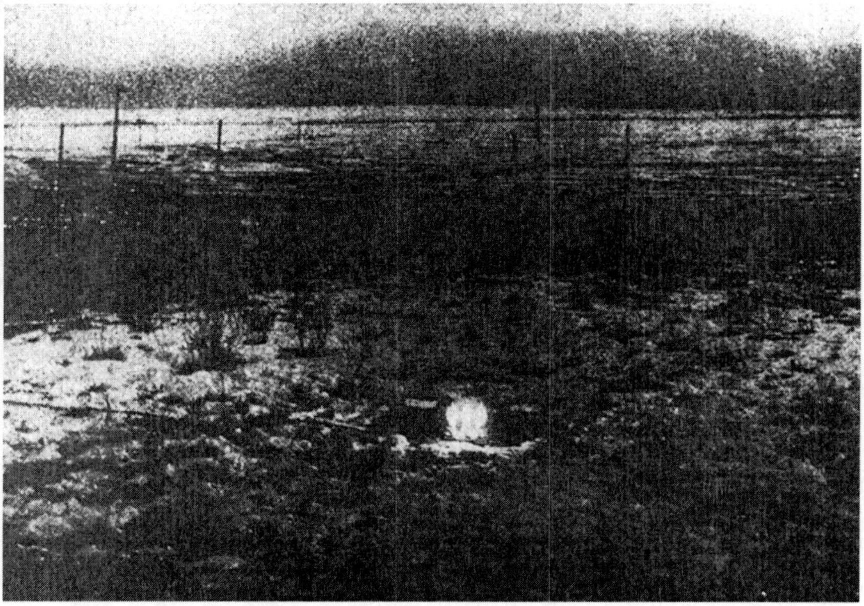

**Fig. 7: Experiment to illustrate an inductive effect of an electrical oscillator of great power.**

I had arrived at the limit of rates obtainable in other ways when the happy idea presented itself to me to resort to the condenser. I arranged such an instrument so as to be charged and discharged alternately in rapid succession through a coil with a few turns of stout wire, forming the primary of a transformer or induction-coil. Each time the condenser was discharged the current would quiver in the primary wire and induce corresponding oscillations in the secondary. Thus, a transformer or induction-coil on new principles was evolved, which I have called "the electrical oscillator," partaking of those unique qualities which characterize the condenser and enabling results to be attained impossible by other means. Electrical effects of any desired character and of intensities undreamed of before are now easily producible by perfected apparatus of this kind, to which frequent reference has been made, and the essential parts of which are shown in Fig. 6. For certain purposes a strong inductive effect is required; for others the greatest possible suddenness; for others again, an exceptionally high rate of vibration or extreme pressure, while for certain other objects immense electrical movements are necessary. The photographs in Figs. 7, 8, 9, and 10, of experiments performed with such an oscillator, may serve to illustrate some of these features and convey an idea of the magnitude of the effects actually produced. The completeness of the titles of the figures referred to makes a further description of them unnecessary.

The photograph shows three ordinary incandescent lamps lighted to full candlepower by currents induced in a local loop consisting of a single wire forming a square of fifty feet each side, which includes the lamps, and which is at a distance of one hundred feet from the primary circuit energized by the oscillator. The loop likewise includes an electrical condenser and is exactly attuned to the vibrations of the oscillator, which is worked at less than five percent of its total capacity.

**Fig. 8: Experiment to illustrate the capacity of the oscillator for producing electrical explosions of great power.**

Note to Fig. 8. The coil, partly shown in the photograph, creates an alternative movement of electricity from the earth into a large reservoir and back at a rate of one hundred thousand alternations per second. The adjustments are such that the reservoir is filled full and bursts at each alternation just at the moment when the electrical pressure reaches the maximum. The discharge escapes with a deafening noise, striking an unconnected coil twenty-two feet away, and creating such a commotion of electricity in the earth that sparks an inch long can be drawn from a water main at a distance of three hundred feet from the laboratory.

The ball shown in the photograph, covered with a polished metallic coating of twenty square feet of surface, represents a large reservoir of electricity, and the inverted tin pan underneath, with a sharp rim, a big opening through which the electricity can escape before filling the reservoir. The quantity of electricity set in movement is so great that, although most of it escapes through the rim of the pan or opening provided, the ball or reservoir is nevertheless alternately emptied and filled to over-flowing (as is evident from the discharge escaping on the top of the ball) one hundred and fifty thousand times per second.

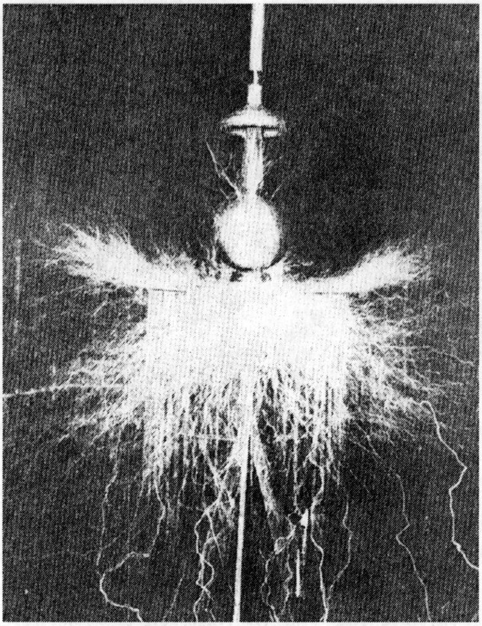

**Fig. 9: Experiment to illustrate the capacity on the oscillator for creating a great electrical movement.**

The discharge, creating a strong draft owing to the heating of the air, is carried upward through the open roof of the building. The greatest width across is nearly seventy feet. The pressure is over twelve million volts, and the current alternates one hundred and thirty thousand times per second.

However extraordinary the results shown may appear, they are but trifling compared with those which are attainable by apparatus designed on these same principles. I have produced electrical discharges the actual path of which, from end to end, was probably more than one hundred feet long; but it would not be difficult to reach lengths one hundred times as great. I have produced electrical movements occurring at the rate of approximately one hundred thousand horsepower, but rates of one, five, or ten million horsepower are easily practicable. In these experiments effects were developed incomparably greater than any ever produced by human agencies, and yet these results are but an embryo of what is to be.

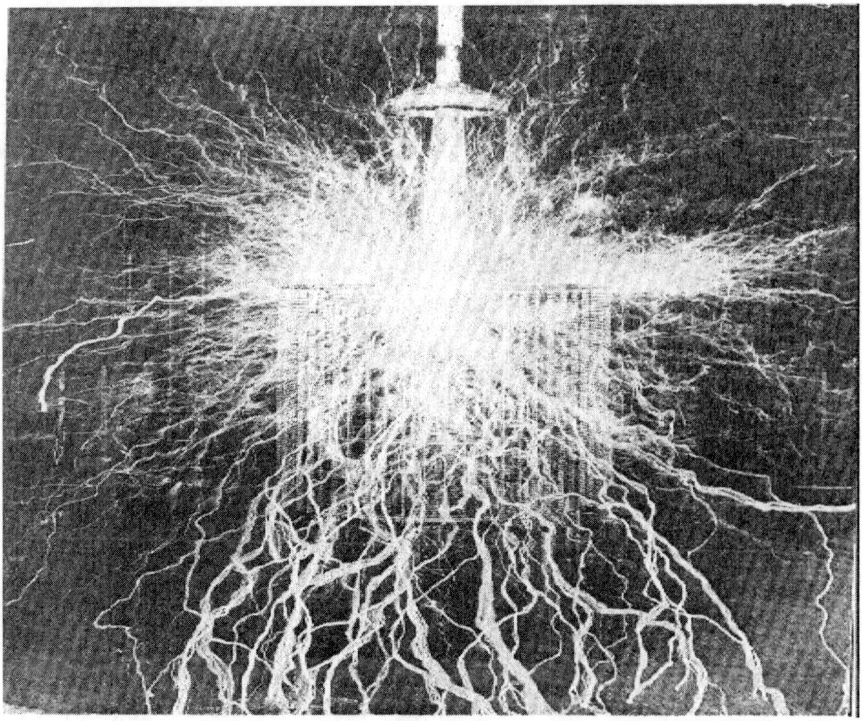

**Fig. 10: Photographic view of an experiment to illustrate an effect of an electrical oscillator delivering energy at a rate of seventy-fife thousand horsepower.**

That communication without wires to any point of the globe is practicable with such apparatus would need no demonstration, but through a discovery which I made I obtained absolute certitude. Popularly explained, it is exactly this: When we raise the voice and hear an echo in reply, we know that the sound of the voice must have reached a distant wall, or boundary, and must have been reflected from the same. Exactly as the sound, so an electrical wave is reflected, and the same evidence which is afforded by an echo is offered by an electrical phenomenon known as a "stationary" wave that is, a wave with fixed nodal and ventral regions. Instead of sending sound vibrations toward a distant wall, I have sent electrical vibrations toward the remote boundaries of the earth, and instead of the wall the earth has replied. In place of an echo, I have obtained a stationary electrical wave, a wave reflected from afar.

Stationary waves in the earth mean something more than mere telegraphy without wires to any distance. They will enable us to attain many important specific results impossible otherwise. For instance, by their use we may produce at will, from a sending-station, an electrical effect in any particular region of the globe; we may determine the relative position or course of a moving object, such as a vessel at sea, the distance traversed by the same, or its speed; or we may send over the earth a wave of electricity traveling at any rate we desire, from the pace of a turtle up to lightning speed.

With these developments we have every reason to anticipate that in a time not very distant most telegraphic messages across the oceans will be transmitted without cables. For short distances we need a "wireless" telephone, which requires no expert operators. The greater the spaces to be bridged, the more rational becomes communication without wires. The cable is not only an easily damaged and costly instrument, but it limits us in the speed of transmission by reason of a certain electrical property inseparable from its construction. A properly designed plant for effecting communication without wires ought to have many times the working capacity of a cable, while it will involve incomparably less expense. Not a long time will pass, I believe, before communication by cable will become obsolete, for not only will signalling by this new method be quicker and cheaper, but also much safer. By using some new means for isolating the messages which I have contrived, an almost perfect privacy can be secured.

I have observed the above effects so far only up to a limited distance of about six hundred miles, but inasmuch as there is virtually no limit to the power of the vibrations producible with such an oscillator, I feel quite confident of the success of such a plant for effecting transoceanic communication. Nor is this all. My measurements and calculations have shown that it is perfectly practicable to produce on our globe, by the use of these principles, an electrical movement of such magnitude that, without the slightest doubt, its effect will be perceptible on some of our nearer planets, as Venus and Mars. Thus, from mere possibility interplanetary communication has entered the stage of probability. In fact, that we can produce a distinct effect on one of these planets in this novel manner, namely, by disturbing the electrical condition of the earth, is beyond any doubt. This way of effecting such communication is, however, essentially different from all others which have so far been proposed by scientific men. In all the previous instances only a minute fraction of the total energy reaching the planet as much as it would be possible to concentrate in a reflector could be utilized by the supposed observer in his instrument. But by the means I have developed he would be enabled to concentrate the larger portion of the entire energy transmitted to the planet in his instrument, and the chances of affecting the latter are thereby increased many millionfold.

Besides machinery for producing vibrations of the required power, we must have delicate means capable of revealing the effects of feeble influences exerted upon the earth. For such purposes, too, I have perfected new methods. By their use we shall likewise be able, among other things, to detect at considerable distance the presence of an iceberg or other object at sea. By their use, also, I have discovered some terrestrial phenomena still unexplained. That we can send a message to a planet is certain, that we can get an answer is probable: man is not the only being in the Infinite gifted with a mind.

# Transmission Of Electrical Energy To Any Distance Without Wires – Now Practicable – The Best Means Of Increasing The Force Accelerating The Human Mass.

The most valuable observation made in the course of these investigations was the extraordinary behaviour of the atmosphere toward electric impulses of excessive electromotive force. The experiments showed that the air at the ordinary pressure became distinctly conducting, and this opened up the wonderful prospect of transmitting large amounts of electrical energy for industrial purposes to great distances without wires, a possibility which, up to that time, was thought of only as a scientific dream. Further investigation revealed the important fact that the conductivity imparted to the air by these electrical impulses of many millions of volts increased very rapidly with the degree of rarefaction, so that air strata at very moderate altitudes, which are easily accessible, offer, to all experimental evidence, a perfect conducting path, better than a copper wire, for currents of this character.

Thus, the discovery of these new properties of the atmosphere not only opened up the possibility of transmitting, without wires, energy in large amounts but, what was still more significant, it afforded the certitude that energy could be transmitted in this manner economically. In this new system it matters little in fact, almost nothing whether the transmission is effected at a distance of a few miles or of a few thousand miles.

While I have not, as yet actually effected a transmission of a considerable amount of energy, such as would be of industrial importance, to a great distance by this new method, I have operated several model plants under exactly the same conditions which will exist in a large plant of this kind, and the practicability of the system is thoroughly demonstrated. The experiments have shown conclusively that, with two terminals maintained at an elevation of not more than thirty thousand to thirty-five thousand feet above sea-level, and with an electrical pressure of fifteen to twenty million volts, the energy of thousands of horsepower can be transmitted over distances which may be hundreds and, if necessary, thousands of miles. I am hopeful, however, that I may be able to reduce very considerably the elevation of the terminals now required, and with this object I am following up an idea which promises such a realization.

There is, of course, a popular prejudice against using an electrical pressure of millions of volts, which may cause sparks to fly at distances of hundreds of feet, but, paradoxical as it may seem, the system, as I have described it in a technical publication, offers greater personal safety than most of the ordinary distribution circuits now used in the cities. This is, in a measure, borne out by the fact that, although I have carried on such experiments for a number of years, no injury has been sustained either by me or any of my assistants.

But to enable a practical introduction of the system, a number of essential requirements are still to be fulfilled. It is not enough to develop appliances by means of which such a transmission can be effected. The machinery must be such as to allow the transformation and transmission, of electrical energy under highly economic and practical conditions. Furthermore, an inducement must be offered to those who are engaged in the industrial exploitation of natural sources of power, as waterfalls, by guaranteeing greater returns on the capital invested than they can secure by local development of the property.

From that moment when it was observed that, contrary to the established opinion, low and easily accessible strata of the atmosphere are capable of conducting electricity, the transmission of electrical energy without wires has become a rational task of the engineer, and one surpassing all others in importance. Its practical consummation would mean that energy would be available for the uses of man at any point of the globe, not in small amounts such as might be derived from the ambient medium by suitable machinery, but in quantities virtually unlimited, from waterfalls. Export of power would then become the chief source of income for many happily situated countries, as the United States, Canada, Central and South America, Switzerland, and Sweden. Men could settle down everywhere, fertilize and irrigate the soil with little effort, and convert barren deserts into gardens, and thus the entire globe could be transformed and made a fitter abode for mankind. It is highly probable that if there are intelligent beings on Mars, they have long ago realized this very idea, which would explain the changes on its surface noted by astronomers. The atmosphere on that planet, being of considerably smaller density than that of the earth, would make the task much more easy.

It is probable that we shall soon have a self-acting heat-engine capable of deriving moderate amounts of energy from the ambient medium. There is also a possibility though a small one that we may obtain electrical energy direct from the sun. This might be the case if the Maxwellian theory is true, according to which electrical vibrations of all rates should emanate from the sun. I am still investigating this subject. Sir William Crookes has shown in his beautiful invention known as the "radiometer" that rays may produce by impact a mechanical effect, and this may lead to some important revelation as to the utilization of the sun's rays in novel ways. Other sources of energy may be opened up, and new methods of deriving energy from the sun discovered, but none of these or similar achievements would equal in importance the transmission of power to any distance through the medium. I can conceive of no technical advance which would tend to unite the various elements of humanity more effectively than this one, or of one which would more add to and more economize human energy. It would be the best means of increasing the force accelerating the human mass. The mere moral influence of such a radical departure would be incalculable. On the other hand, if at any point of the globe energy can be obtained in limited quantities from the ambient medium by means of a self-acting heat-engine or otherwise, the conditions will remain the same as before. Human performance will be increased, but men will remain strangers as they were.

I anticipate that any, unprepared for these results, which, through long familiarity, appear to me simple and obvious, will consider them still far from practical application. Such reserve, and even opposition, of some is as useful a quality and as necessary an element in human progress as the quick receptivity and enthusiasm of others. Thus, a mass which resists the force at first, once set in movement, adds to the energy. The scientific man does not aim at an immediate result. He does not expect that his advanced ideas will be readily taken up. His work is like that of the planter for the future. His duty is to lay the foundation for those who are to come and point the way. He lives and labours and hopes with the poet who says:

Schaff' das Tagwerk meiner Hände,

Hohes Glück, dass ich's vollende!

Laß, o laß mich nicht ermatten!

Nein, es sind nicht leere Träume:

Jetzt nur Stangen, diese Bäume

Geben einst noch Frucht und Schatten. [1]

1 Daily work my hands' employment,

To complete is pure enjoyment!

Let, oh, let me never falter!

No! there is no empty dreaming:

No! these trees, but bare poles seeming,

Yet will yield both food and shelter!

Goethe's "Hope"

Translated by William Gibson, Com. U. S. N.

## LIST OF TITLES WITH ISBN NO.

| ISBN | TITLE |
|---|---|
| 9788194914129 | 1984 |
| 9789390575220 | 1984 & Animal Farm (2In1) |
| 9789390575572 | 1984 & Animal Farm (2In1): The International Best-Selling Classics |
| 9789390575848 | 35 Sonnets |
| 9789390575329 | A Clergyman's Daughter |
| 9789390575923 | A Study In Scarlet |
| 9789390896097 | A Tale Of Two Cities |
| 9789390896837 | Abide in Christ |
| 9789390896202 | Abraham Lincoln |
| 9789390896912 | Absolute Surrender |
| 9789390896608 | African American Classic Collection |
| 9789390575305 | Aldous Huxley: The Collected Works |
| 9789390896141 | An Autobiography of M. K. Gandhi |
| 9789390575886 | Animal Farm |
| 9789390575619 | Animal Farm & The Great Gatsby (2In1) |
| 9789390575626 | Animal Farm & We |
| 9789390896158 | Anna Karenina |
| 9789390575534 | Antic Hay |
| 9789390896165 | Antony & Cleopatra |
| 9789390896172 | As I Lay Dying |
| 9789390896226 | As You like it |
| 9789390575671 | At Your Command |
| 9789390575350 | Awakened Imagination |
| 9789390575114 | Be What You Wish |
| 9789390896233 | Believe In yourself |
| 9789390896998 | Best of Charles Darwin: The Origin of Species & Autobiography |
| 9789390896684 | Best Of Horror : Dracula And Frankenstein |
| 9789390575503 | Best Of Mark Twain (The Adventures of Tom Sawyer AND The Adventures of Huckleberry Finn) |
| 9789390896769 | Black History Collection |
| 9789390575756 | Brave New World, Animal Farm & 1984 (3in1) |

| | |
|---|---|
| 9789390896240 | Brother Karamzov |
| 9789390575053 | Bulleh Shah Poetry |
| 9789390575725 | Burmese Days |
| 9789390896257 | Bushido |
| 9789390896066 | Can't Hurt Me |
| 9788194914112 | Chanakya Neeti: With The Complete Sutras |
| 9789390896042 | Crime and Punishment |
| 9789390575527 | Crome Yellow |
| 9789390575046 | Down and Out in Paris and London |
| 9789390896844 | Dracula |
| 9789390575442 | Emersons Essays: The Complete First & Second Series (Self-Reliance & Other Essays) |
| 9789390575749 | Emma |
| 9789390575817 | Essential Tozer Collection - The Pursuit of God & The Purpose of Man |
| 9789390896578 | Fascism What It Is and How to Fight It |
| 9789390575688 | Feeling is the Secret |
| 9789390575190 | Five Lessons |
| 9789390575954 | Frankenstein |
| 9789390575237 | Franz Kafka: Collected Works |
| 9789390575282 | Franz Kafka: Short Stories |
| 9789390575060 | George Orwell Collected Works |
| 9789390575077 | George Orwell Essays |
| 9789390575213 | George Orwell Poems |
| 9788194914150 | Greatest Poetry Ever Written Vol 1 |
| 9788194914143 | Greatest Poetry Ever Written Vol 1 |
| 9789390896301 | Gulliver's Travel |
| 9789390575961 | Gunaho Ka Devta |
| 9789390575893 | H. P. Lovecraft Selected Stories Vol 1 |
| 9789390575978 | H. P. Lovecraft Selected Stories Vol 2 |
| 9789390896059 | Hamlet |
| 9789390575022 | His Last Bow: Some Reminiscences of Sherlock Holmes |
| 9789390896134 | History of Western Philosophy |
| 9789390575121 | Homage To Catalonia |

| | |
|---|---|
| 9789390896219 | How to develop self-confidence and Improve public Speaking |
| 9789390896295 | How to enjoy your life and your Job |
| 9789390575633 | How to own your own mind |
| 9789390896318 | How to read Human Nature |
| 9789390896325 | How to sell your way through the life |
| 9789390896370 | How to use the laws of mind |
| 9789390896387 | How to use the power of prayer |
| 9789390896028 | How to win friends & Influence People |
| 9788194824176 | How To Win Friends and Influence People |
| 9789390896103 | Humility The Beauty of Holiness |
| 9789390896653 | Imperialism the Highest Stage of Capitalism |
| 9789390575084 | In Our Time |
| 9789390575169 | In Our Time & Three Stories and Ten poems |
| 9789390575145 | James Allen: The Collected Works |
| 9789390896189 | Jesus Himself |
| 9789390575480 | Jo's Boys |
| 9789390896394 | Julius Caesar |
| 9789390575404 | Keep the Aspidistra Flying |
| 9789390896400 | Kidnapped |
| 9789390896424 | King Lear |
| 9789390575824 | Lady Susan |
| 9789390896455 | Law of Success |
| 9789390896264 | Lincoln The Unknown |
| 9789390575565 | Little Men |
| 9789390575640 | Little Women |
| 9788194914174 | Lost Horizon |
| 9789390896462 | Macbeth |
| 9789390896929 | Man Eaters of Kumaon |
| 9789390896523 | Man The Dwelling Place of God |
| 9789390896349 | Man The Dwelling Place of God |
| 9789390575909 | Mansfield Park |
| 9788194914136 | Manto Ki 25 Sarvshreshth Kahaniya |
| 9789390896509 | Marxism, Anarchism, Communism |
| 9789390575664 | Mathematical Principles of Natural Philosophy |

| | |
|---|---|
| 9788194914198 | Meditations |
| 9789390575800 | Mein Kampf |
| 9789390575794 | Memory How To Develop, Train, And Use It |
| 9789390896486 | Mind Power |
| 9789390896585 | Money |
| 9789390575039 | Mortal Coils |
| 9789390575770 | My Life and Work |
| 9789390896035 | Narrative of the Life of Frederick Douglass |
| 9789390575152 | Neville Goddard: The Collected Works |
| 9789390575985 | Northanger Abbey |
| 9789390896530 | Notes From Underground |
| 9789390896547 | Oliver Twist |
| 9789390575459 | On War |
| 9789390575541 | One, None and a Hundred Thousand |
| 9789390896554 | Othelo |
| 9789390575435 | Out Of This World |
| 9789390575015 | Persuasion |
| 9789390575510 | Prayer The Art Of Believing |
| 9789390575091 | Pride and Prejudice |
| 9789390896561 | Psychic Perception |
| 9789390575381 | Rabindranath Tagore - 5 Best Short Stories Vol 2 |
| 9789390575367 | Rabindranath Tagore - Short Stories (Masters Collections Including The Childs Return) |
| 9789390575374 | Rabindranath Tagore 5 Best Short Stories Vol 1 (Including The Childs Return |
| 9789390896622 | Romeo & Juliet |
| 9789390896127 | Sanatana Dharma |
| 9789390575596 | Seedtime & Harvest |
| 9789390896639 | Selected Stories of Guy De Maupassant |
| 9789390575206 | Self-Reliance & Other Essays |
| 9789390575176 | Sense and Sensibility |
| 9789390575299 | Shyamchi Aai |
| 9789390896738 | Socialism Utopian and Scientific |
| 9789390896646 | Success Through a Positive Mental Attitude |
| 9789390575428 | The Adventures of Huckleberry Finn |

| | |
|---|---|
| 9789390575183 | The Adventures of Sherlock Holmes |
| 9789390575343 | The Adventures of Tom Sawyer |
| 9789390896691 | The Alchemy Of Happiness |
| 9789390575862 | The Art Of Public Speaking |
| 9789390896288 | The Autobiography Of Charles Darwin |
| 9788194914181 | The Best of Franz Kafka: The Metamorphosis & The Trial |
| 9789390575008 | The Call Of Cthulhu and Other Weird Tales |
| 9789390575107 | The Case-Book of Sherlock Holmes |
| 9789390896110 | The Castle Of Otranto |
| 9789390896745 | The Communist Manifesto |
| 9789390575589 | The Complete Fiction of H. P. Lovecraft |
| 9789390575497 | The Complete Works of Florence Scovel Shinn |
| 9789390896820 | The Conquest of Breard |
| 9789390896813 | The Diary of a Young Girl |
| 9789390896332 | The Diary of a Young Girl The Definitive Edition of the Worlds Most Famous Diary |
| 9789390575701 | The Great Gatsby, Animal Farm & 1984 (3In1) |
| 9789390575312 | The Greatest Works Of George Orwell (5 Books) Including 1984 & Non-Fiction |
| 9789390575992 | The Hound of Baskervilles |
| 9789390896707 | The Idiot |
| 9789390896714 | The Invisible Man |
| 9789390575657 | The Knowledge of the holy |
| 9789390575558 | The Law & the Promise |
| 9789390896721 | The Law Of Attraction |
| 9789390896776 | The Leader in you |
| 9789390896363 | The Life of Christ |
| 9789390896196 | The Man-Eating Leopard of Rudraprayag |
| 9789390896783 | The Master Key to Riches |
| 9789390575268 | The Memoirs Of Sherlock Holmes |
| 9789390896479 | The Midsummer Night's Dream |
| 9789390575466 | The Mill On The Floss |
| 9789390896790 | The Miracles of your mind |
| 9789390896660 | The Mutual Aid A Factor in Evolution |
| 9789390896448 | The Origin of Species |

| | |
|---|---|
| 9789390896905 | The Peter Kropotkin Anthology The Conquest of Bread & Mutual Aid A Factor of Evolution |
| 9789390896806 | The Picture of Dorian Gray |
| 9789390896271 | The Picture of Dorian Gray |
| 9789390575275 | The Power Of Awareness |
| 9789390896356 | The Power of Concentration |
| 9788194824169 | The Power of Positive Thinking |
| 9789390575411 | The Power of the Spoken Word |
| 9788194914105 | The Power Of Your Subconscious Mind |
| 9789390896899 | The Power of Your Subconscious Mind |
| 9789390896417 | The Principles of Communism |
| 9789390575787 | The Psychology Of Mans Possible Evolution |
| 9789390896615 | The Psychology of Salesmanship |
| 9789390575732 | The Pursuit of God |
| 9789390575398 | The Pursuit of Happiness |
| 9789390896851 | The Quick and Easy Way to effective Speaking |
| 9789390575947 | The Return Of Sherlock Holmes |
| 9789390575138 | The Road To Wigan Pier |
| 9789390896981 | The Root of the Righteous |
| 9789390575855 | The Science Of Being Well |
| 9788194914167 | The Science Of Getting Rich, The Science Of Being Great & The Science Of Being Well (3In1) |
| 9789390896011 | The Screwtape Letters |
| 9789390896073 | The Screwtape Letters |
| 9789390575336 | The Secret Door to Success |
| 9789390575695 | The Secret Of Imagining |
| 9789390896868 | The Secret Of Success |
| 9789390896431 | The Seven Last Words |
| 9789390575930 | The Sign of the Four |
| 9789390896004 | The Sonnets |
| 9789390896516 | The Souls of Black Folk |
| 9789390896875 | The Sound and The Fury |
| 9789390575244 | The State and Revolution |
| 9789390896882 | The Story of My Life |
| 9789390896936 | The Story Of Oriental Philosophy |

| | |
|---|---|
| 9789390896752 | The Strange Case of Dr. Jekyll and Mr. Hyde |
| 9789390896943 | The Tempest |
| 9789390575916 | The Valley Of Fear |
| 9789390575879 | The Wind in the willows |
| 9789390896080 | The Wind in the willows |
| 9789390575763 | Their eyes were watching gofd |
| 9789390575831 | Three Stories |
| 9789390896950 | Twelfth Night |
| 9789390896592 | Twelve Years a Slave |
| 9789390896677 | Up from Slavery |
| 9789390896974 | Value Price and Profit |
| 9789390896967 | Wake Up and Live |
| 9789390896493 | With Christ in the School of Prayer |
| 9789390575602 | Your Faith is Your Fortune |
| 9789390575473 | Your Infinite Power To Be Rich |
| 9789390575251 | Your Word is Your Wand |
| 9789390575718 | Youth |
| 9789391316099 | A Christmas Carol |
| 9789391316105 | A Doll's House |
| 9789391316501 | A Passage to India |
| 9789391316709 | A Portrait of the Artist as a Young Man |
| 9789391316112 | A Tale of Two Cities |
| 9789391316747 | A Tear and a Smile |
| 9789391316167 | Agnes Gray |
| 9789391316174 | Alice's Adventures in Wonderland |
| 9789391316136 | Anandamath |
| 9789391316181 | Anne Of Green Gables |
| 9789391316754 | Anthem |
| 9789391316198 | Around The World in 80 Days |
| 9789391316013 | As A Man Thinketh |
| 9789391316242 | Autobiography of a Yogi |
| 9789391316266 | Beyond Good and Evil |
| 9789391316761 | Bleak House |
| 9789391316778 | Chitra, a Play in One Act |
| 9789391316310 | David Copperfield |

| 9789391316075 | Demian |
|---|---|
| 9789391316785 | Dubliners |
| 9789391316051 | Favourite Tales from the Arabian Nights |
| 9789391316235 | Gitanjali |
| 9789391316068 | Gravity |
| 9789391316150 | Great Speeches of Abraham Lincoln |
| 9789391316662 | Guerilla Warfare |
| 9789391316839 | Kim |
| 9789391316822 | Mother |
| 9789391316211 | My Childhood |
| 9789391316846 | Nationalism |
| 9789391316327 | Oliver Twist |
| 9789391316853 | Pygmalion |
| 9789391316334 | Relativity: The Special and the General Theory |
| 9789391316389 | Scientific Healing Affirmation |
| 9789391316341 | Sons and Lovers |
| 9789391316587 | Tales from India |
| 9789391316372 | Tess of The D'Urbervilles |
| 9789391316396 | The Awakening and Selected Stories |
| 9789391316402 | The Bhagvad Gita |
| 9789391316303 | The Book of Enoch |
| 9789391316228 | The Canterville Ghost |
| 9789391316907 | The Dynamic Laws of Prosperity |
| 9789391316006 | The Great Gatsby |
| 9789391316860 | The Hungry Stones and Other Stories |
| 9789391316433 | The Idiot |
| 9789391316440 | The Importance of Being Earnest |
| 9789391316297 | The Light of Asia |
| 9789391316914 | The Madman His Parables and Poems |
| 9789391316457 | The Odyssey |
| 9789391316921 | The Picture of Dorian Gray |
| 9789391316464 | The Prince |
| 9789391316938 | The Prophet |
| 9789391316945 | The Republic |
| 9789391316518 | The Scarlet Letter |

| | |
|---|---|
| 9789391316143 | The Seven Laws of Teaching |
| 9789391316525 | The Story of My Experiments with Truth |
| 9789391316532 | The Tales of the Mother Goose |
| 9789391316549 | The Thirty Nine Steps |
| 9789391316594 | The Time Machine |
| 9789391316600 | The Turn of the Screw |
| 9789391316983 | The Upanishads |
| 9789391316617 | The Yellow Wallpaper |
| 9789391316426 | The Yoga Sutras of Patanjali |
| 9789391316990 | Ulysses |
| 9789391316624 | Utopia |
| 9789391316679 | Vanity Fair |
| 9789391316020 | What Is To Be Done |
| 9789391316686 | Within A Budding Grove |
| 9789391316693 | Women in Love |